I0477785

Optimized Living
The Mathematics of Habits and Productivity

(Math Can Elevate Your Habits, Refine Your Decisions, and Optimize Your Time)

By

Dr. Jitendra Singh

Associate Professor

Department of Mathematics,
Raj Rishi College, Alwar (Raj.), India, 301001

Editor
Mrs. Preety Verma

ISBN: 979-8-3458-3232-5
Amazon Kindle Direct Publishing, Columbia, SC USA
https://amazon.com

Preface

In a world where efficiency and productivity often feel just out of reach, mathematics can offer practical solutions for achieving a well-balanced life. While math may traditionally evoke images of complex equations or abstract theorems, its principles are profoundly relevant to everyday decisions, personal habits, and time management. This book, Mathematics for Personal Habits and Productivity: How Math Can Enhance Your Habits, Decisions, and Time Management, explores the surprising ways that numbers and simple calculations can unlock greater control over our routines, enhance decision-making, and clarify our goals. Here, we delve into the fundamentals of quantifying habits, applying logical frameworks to decisions, and using mathematical models to optimize time. From calculating small steps to achieve a big goal to analyzing personal data for more productive routines, each chapter translates mathematical concepts into tools for measurable growth and self-improvement. Whether you're a student, a professional, or anyone seeking to lead a more purposeful, effective life, this book offers a path to achieving those goals—one calculation at a time.

Dr. Jitendra Singh
November 2024

Contents

12 Conclusion 119

Chapter 1

Introduction to Mathematical Thinking

This chapter introduces the use of mathematical thinking in self-improvement by turning personal goals into measurable and trackable objectives. It emphasizes setting specific, quantifiable goals, using tools for tracking progress, breaking large goals into manageable steps, and adjusting goals based on regular reviews. Celebrating small wins is highlighted as essential for motivation. By applying cumulative tracking, trend analysis, and probability, this approach transforms abstract goals into structured, data-driven strategies for productivity and habit-building, empowering readers to make informed adjustments and maintain steady progress.

1.1 Overview of Personal Habits and Productivity

In this Section, we explore the foundation of mathematical thinking and its applications in enhancing personal habits and productivity. Mathematics, often seen as a tool for science and engineering, can equally provide a structured framework to analyze and improve daily routines, set achievable goals, and make better decisions.

1.1.1 Mathematical Thinking for Self Improvement

Mathematical thinking involves applying logic, structure, and quantitative analysis to problem-solving. It is a systematic approach to understanding complex issues by breaking them down into simpler, manageable parts. By viewing personal habits through a mathematical lens, we can objectively analyze our progress, make data-driven decisions, and set realistic expectations.

For example, if you want to improve your productivity, you can break down your daily tasks into manageable chunks (e.g., using time-blocking techniques) and measure how long each task takes. This allows you to identify where you are spending too much time and adjust accordingly. Another example is tracking your eating habits by quantifying your daily calorie intake, the nutritional value of meals, and portion sizes. This allows you to make improvements based on hard data rather than subjective observations. Similarly, for exercise habits, you can measure the number of steps walked daily or the minutes spent exercising per week to create a routine that supports gradual improvement.

1.1.2 The Role of Quantification in Self Improvement

Quantifying our actions and habits allows us to see tangible progress over time. This may involve tracking daily routines, measuring productivity, or analyzing habits with numbers. Awareness is the first step to change. Quantitative measures help us recognize patterns, such as peak productivity times or common distractions, and make adjustments accordingly.

For instance, you can use a habit-tracking app to monitor how many hours you spend on productive activities versus distractions. Seeing your time allocations in graphs can help identify where you can improve. Another example is tracking the number of hours you sleep each night and comparing this data with your alertness and performance the next day. If you notice a correlation between insufficient sleep and decreased productivity, you can adjust your routine. Additionally, you can use a simple spreadsheet to track your budget by categorizing expenses and monitoring monthly trends. This will allow you to identify areas where you're overspending and adjust accordingly.

1.1.3   Setting SMART Goals and Mathematics

The SMART goal framework provides a structured approach to setting effective goals. Mathematics aids in defining measurable and achievable targets within specified time frames. By breaking down goals into numbers (e.g., a weekly study quota or daily exercise minutes), we create a clear path to success, using math as a guiding tool.

If your goal is to read more books, break this into measurable targets such as "Read 10 pages per day." In one month, you will have read approximately 300 pages, which you can measure against your goal. Similarly, if you're working on weight loss, you can set a goal of losing 1-2 pounds per week. This gives you a tangible target and allows you to track progress weekly. Another example for improving your physical fitness could involve setting a goal of exercising 30 minutes per day, 5 times per week, and tracking your consistency over time to see if adjustments are needed.

1.1.4   Understanding the Growth Mindset

A growth mindset embraces continuous improvement and learning, often fostered by feedback loops. Mathematics supports this by allowing for measurable incremental growth. Using metrics like percentage improvements or average increases over time, mathematics shows how small steps contribute to larger accomplishments, reinforcing a growth mindset.

If you're learning a new skill, such as playing an instrument, you can track how many minutes of practice you do each day. Even small increases in practice time will compound over weeks and months. In language learning, measuring the number of new words learned per week or the hours spent practicing pronunciation helps reinforce the concept that continuous effort leads to improvement. For a professional goal like writing a book, tracking your word count daily or weekly can help you see measurable progress. Even writing just 500 words per day will result in substantial progress over time.

Feedback Loops and Adjustments Using Data

Tracking data over time allows us to see where adjustments are needed, reinforcing actions that work and adapting those that don't. Positive reinforcement occurs when progress is acknowledged, which mathematics can quantify, fostering motivation and commitment to habit changes.

If you're trying to improve your diet, tracking your food choices and measuring how you feel each day can show patterns. For instance, you may notice higher energy levels on days when you eat more vegetables, encouraging you to make better food choices. If you're working on a professional project, using progress bars or tracking milestones can provide constant visual feedback, encouraging continued effort towards your goals. You can also adjust your fitness goals by tracking performance metrics like speed, strength, or stamina. If you're not meeting your targets, data allows you to reassess your workout plan and make necessary changes.

1.2   The Basics of Quantifying and Tracking Habits

Quantifying and tracking habits is one of the most effective ways to gain control over personal improvement. By measuring habits, we make our routines more transparent and can clearly see where to make adjustments for better productivity, health, or any other area of life. This process involves recording specific data points related to a habit and analyzing this information to identify trends, progress, or areas for change.

1.2.1   Defining the Habits You Want to Track

Before you can quantify and track a habit, you must first clearly define the habit itself. A habit can be any behavior or action that is repeated regularly, but for tracking purposes, it is best to make it specific, measurable, and realistic.

Example 1: Tracking Exercise For instance, if your goal is to improve physical fitness, you can define the habit as "exercise 30 minutes per day." This gives you a specific, measurable action to track. The frequency, duration, and type of exercise can all be recorded in a systematic way, allowing you to monitor your progress over time.

Example 2: Tracking Sleep Another example could be improving your sleep. Instead of a vague goal like "get better sleep," you could define your habit as "sleep at least 7 hours per night." This is both specific and measurable. By tracking your sleep duration every night, you can start to see patterns, such as how sleep affects your energy levels the next day.

Example 3: Tracking Eating Habits Similarly, you might want to track your eating habits. You could define the habit as "eat three balanced meals per day," where you could monitor the types of food you consume, the timing, and portion sizes. Using a food journal or an app can help quantify this habit and ensure consistency.

1.2.2   Selecting the Right Tools for Tracking

To quantify and track habits effectively, you need the right tools. These could range from simple paper journals to high-tech mobile apps, depending on your preference and needs. Here are some common tools:

Example 1: Habit-Tracking Apps Apps such as Habitica or Streaks allow you to log habits daily and track progress with visual feedback, such as streaks of consecutive successful days. They often offer reminders, rewards, and other motivational features to help stay on track.

Example 2: Spreadsheets or Journals For those who prefer a low-tech approach, using a spreadsheet to track habits can be very effective. In a spreadsheet, you can create a table with each habit listed as a column and the days of the month as rows. Simply mark off each day that you complete the habit, and this can give you a clear visual representation of your progress.

Example 3: Wearable Devices For more detailed tracking, wearable devices like Fitbits or Apple Watches can track metrics like steps, calories burned, and even sleep quality. These devices can provide automatic data logging, which makes it easier to track habits without having to manually enter information.

1.2.3   Setting Realistic Goals for Habit Tracking

Once you've defined the habit and selected your tracking tools, the next step is setting realistic goals for how often you want to engage in the habit and the level of performance expected. Goals should be challenging but attainable and should allow for growth over time.

Example 1: Incremental Goals for Exercise If you're just starting with exercise, a realistic goal might be "exercise 20 minutes per day for the first week." Once this becomes a habit, you can incrementally increase the duration to 30 minutes or more. Over time, your goal becomes more challenging, but your tracking data will show steady improvement.

Example 2: Progressive Sleep Goals For sleep, you might start with a goal to get "at least 6 hours of sleep each night." Once you consistently meet that goal, you can increase your target to 7 hours, and later 8 hours. This approach allows you to build up to healthier habits gradually without feeling overwhelmed.

Example 3: Small Steps for Eating Habits For eating habits, start small. You could aim to "eat vegetables with at least two meals a day" or "reduce sugar intake to 25 grams per day." Over time, these goals can become more specific, such as "eat five servings of vegetables per day" or "eliminate added sugars entirely."

1.2.4   Tracking Your Progress and Identifying Patterns

Once you've set your goals and started tracking your habits, it's crucial to regularly evaluate your progress. This can be done by reviewing your tracking tool (app, spreadsheet, journal, etc.) to identify trends and patterns in your behavior. Are there certain times of the day when you struggle to complete your habits? Are there specific obstacles or distractions that prevent you from following through? Understanding these patterns can help you make informed adjustments to improve consistency.

Example 1: Reviewing Exercise Consistency If you've been tracking exercise, you might notice that you consistently miss your goal on weekends. This could be due to a lack of structure or planning. Recognizing this trend will allow you to make adjustments, such as setting a specific time for exercise on weekends or preparing in advance to overcome obstacles like social events or family obligations.

Example 2: Sleep Tracking Insights In your sleep data, you may observe that you sleep less on days when you drink coffee in the afternoon. This pattern can prompt you to eliminate caffeine in the later hours of the day, which may lead to improved sleep quality over time.

Example 3: Adjusting Eating Habits If you track your eating habits, you may find that you tend to snack on unhealthy foods in the evening. By recognizing this pattern, you can take proactive steps such as preparing healthier snacks or setting a rule not to eat after a certain time.

1.2.5   Adapting and Improving Habits

The most powerful benefit of tracking habits is the ability to adapt and improve based on data-driven insights. When you consistently track habits over time, the data will give you concrete feedback on what is working and what is not. Use this information to make adjustments, optimize your routines, and ensure that your habits continue to serve your personal goals.

Example 1: Improving Exercise with Data If you find that certain types of exercise are more enjoyable and yield better results, you can adjust your workout routine to focus on those activities. If running is causing discomfort or you aren't seeing progress, you might switch to cycling or swimming and track the results over time.

Example 2: Improving Sleep Hygiene By reviewing your sleep data and noticing that certain habits lead to better rest (such as no screen time before bed), you can build those habits into your evening routine. Adjusting factors like lighting, temperature, and bedtime consistency can also help improve sleep quality.

Example 3: Tailoring Your Diet Data from habit tracking might reveal that you feel better when you eat more plant-based meals. Armed with this information, you can adjust your meal planning, increasing the number of plant-based meals and tracking the impact on your energy levels and health.

1.3 Setting Measurable Goals

Setting measurable goals and establishing a framework for progress are key to turning abstract desires into achievable actions. This process allows you to track your success, adjust strategies when needed, and stay motivated over time. By ensuring that your goals are measurable, you can make objective assessments about your growth and areas for improvement.

1.3.1 Defining Measurable Goals

Measurable goals are essential because they provide concrete targets that can be tracked over time. Instead of vague goals like "get better at writing" or "exercise more," measurable goals should include specific criteria that allow you to evaluate your success. These could involve quantity, time, frequency, or specific outcomes.

Example 1: Writing Goal Rather than setting a goal like "write more," you could define a measurable goal such as "write 500 words every day."

This goal is specific (500 words), measurable (you can count the words), and time-bound (daily). Tracking this goal allows you to measure progress each day, providing both motivation and clear results.

Example 2: Exercise Goal A more measurable exercise goal could be "run 3 times a week for 30 minutes." This goal specifies frequency (3 times a week), duration (30 minutes), and type of activity (running), making it easy to track and adjust as necessary.

Example 3: Reading Goal For reading, a measurable goal could be "read 10 pages every day." This goal is easy to track, and you can quickly assess whether you are meeting your target. Over time, this goal helps build a reading habit with tangible progress.

1.3.2 Establishing a Framework for Tracking Progress

Once you have defined measurable goals, the next step is to establish a system or framework for tracking your progress. This framework can involve tools, routines, or habits that support goal achievement. Having a structured system in place ensures that progress is continuously evaluated, allowing for timely adjustments when necessary.

Example 1: Habit-Tracking App A popular tool for tracking progress is a habit-tracking app, such as Habitica or Streaks. These apps allow you to input your goals, track them daily, and visually monitor your progress with features like streaks or achievement badges. This provides instant feedback, reinforcing consistency.

Example 2: Weekly Review System A more manual approach could be a weekly review system. For example, you could set aside time every Sunday to review the goals you set for the week, check your progress, and make any necessary adjustments. This can involve a simple chart or a more detailed journal entry to track the metrics of each goal.

Example 3: Fitness Tracker For fitness goals, wearable devices such as a Fitbit or Apple Watch can track metrics like steps, heart rate, and calories burned. These devices give you real-time feedback and cumulative progress towards your goal, making it easier to stay on track.

1.3.3   Breaking Down Goals into Manageable Steps

Breaking your goals into smaller, manageable steps is essential to make them less overwhelming and more achievable. Each small step should build toward the larger goal, providing continuous milestones that motivate you along the way.

Example 1: Writing Goal Breakdown If your goal is to write a book, breaking it down into manageable steps might involve setting daily targets, such as "write 500 words a day" or "write 5 pages each week." Each page or word count serves as a mini-milestone, making the larger task feel more achievable.

Example 2: Fitness Goal Breakdown For fitness, you might break your exercise goal into smaller milestones, such as "run 1 mile the first week," "run 2 miles the second week," and so on. This way, you gradually increase the intensity, avoiding burnout and promoting consistency.

Example 3: Learning a New Skill If you are learning a new skill, such as playing the guitar, you could set smaller, incremental goals like "learn one chord per day" or "practice for 20 minutes every day." This allows you to make continuous progress without feeling overwhelmed by the larger goal of becoming proficient in the skill.

1.3.4   Regularly Reviewing and Adjusting Goals

Progress isn't always linear, and it's essential to regularly review your goals to assess if they are still realistic or need adjustment. If you're falling behind or achieving the goal too easily, making adjustments can ensure that the goal remains challenging but attainable. Review helps identify areas of improvement or even refine your goals as circumstances change.

Example 1: Reviewing Fitness Goals If your original goal was to "run 3 times a week for 30 minutes," but you find it too easy or too hard, you can adjust the goal. Perhaps you could increase the time to 40 minutes or change the goal to a more specific challenge, like running 10 kilometers within a month.

Example 2: Adjusting a Writing Goal If you set a goal to "write 500 words every day," but find that certain days are harder to meet this goal, you can adjust the number of words on busy days or focus on writing 3 days a week instead. This helps keep the goal achievable while still progressing towards the larger objective.

Example 3: Review Learning Goals If you are learning a new language, you might set a goal to "learn 10 new words every week." If you're progressing faster, you might increase the goal to 15 words a week. If it's too difficult, you could reduce it to 5 words per week. The key is to adapt the goal based on progress.

1.3.5 Celebrating Small Wins

Celebrating small wins is an important part of maintaining motivation. Acknowledge each small milestone you reach on your journey towards a larger goal. By doing so, you build momentum, which makes the overall process more enjoyable and sustainable.

Example 1: Writing Progress Celebration If you complete a set number of words or chapters, take time to celebrate. This could involve taking a break, sharing your progress with a friend, or treating yourself to something small that reinforces your accomplishment.

Example 2: Fitness Celebration After meeting a fitness milestone, such as running 5 kilometers for the first time, treat yourself to a special reward, like a new pair of running shoes or a relaxing activity like a massage.

Example 3: Language Learning Celebration If you've consistently learned 100 new words in your language learning goal, celebrate by watching a movie in the target language or having a conversation with a fluent speaker. This reinforces the habit and gives a sense of achievement.

Chapter 2

The Science of Habit Formation: A Quantitative Approach

In this chapter, we explore how mathematical principles can be applied to understand and reinforce habits. It covers key insights from habit science, including habit loops and triggers, and delves into the mathematics of habit formation, such as exponential growth, compounding effects, and the impact of small wins. Practical tools for tracking habits, like habit streaks and percentage-based goal achievement, are also discussed, providing readers with concrete methods for measuring and maintaining progress. This chapter offers a foundation for building productive habits through a structured, mathematical approach.

2.1   Key Insights from Habit

Understanding the science behind habits, including the concepts of habit loops and triggers, can help in systematically creating and reinforcing desired behaviors. By applying mathematical modeling, we can quantify and predict habit formation patterns over time, allowing for data-driven improvements in our routines.

2.1.1   The Habit Loop: Cue, Routine, and Reward

The habit loop, a concept introduced by habit scientists, consists of three components: - **Cue (or Trigger)**: The event or signal that initiates the habit. - **Routine**: The behavior or action taken in response to the cue. - **Reward**: The benefit received, reinforcing the habit loop.

By modeling this loop mathematically, we can examine the probability of completing a habit cycle and analyze how modifications to each component affect habit strength.

Example 1: Modeling Habit Probability Suppose $P(H)$ represents the probability of completing a habit on any given day. This probability depends on the strength of the cue $P(C)$, routine $P(R)$, and reward $P(W)$:

$$P(H) = P(C) \times P(R) \times P(W)$$

If each component has an initial success probability of 0.8, then the likelihood of completing the habit is:

$$P(H) = 0.8 \times 0.8 \times 0.8 = 0.512$$

This model suggests that strengthening any of the components increases the habit completion probability, and even small changes can significantly impact consistency over time.

Example 2: Analyzing Cue Effectiveness To increase the success rate of a habit, one could focus on the cue. For instance, setting a reminder or placing visual cues (e.g., gym clothes by the door) could improve the probability of the cue $P(C)$ being noticed. If $P(C)$ increases from 0.8 to 0.9, the overall habit success probability becomes:

$$P(H) = 0.9 \times 0.8 \times 0.8 = 0.576$$

This shows that enhancing cues alone can lead to a higher likelihood of habit formation.

2.1.2 Quantifying Habit Strength

Habit formation depends on the consistency and frequency of performing the habit. By tracking these factors over time, we can mathematically analyze habit strength and persistence.

Example 1: Consistency Rate Let f represent the frequency of performing a habit per week (e.g., $f = 5$ days). Define $C(n)$ as the consistency rate over n weeks:

$$C(n) = \frac{\text{Total Days Performed in } n \text{ Weeks}}{7n}$$

If the habit was performed on 30 out of 35 possible days, then the consistency rate over 5 weeks is:

$$C(5) = \frac{30}{35} = 0.857$$

A high consistency rate, closer to 1, indicates strong habit formation.

Example 2: Exponential Habit Growth Model Assume that each time the habit is completed, it reinforces the routine, increasing the likelihood of repeating it by a small growth factor r. Over time, the habit strength $S(t)$ can be modeled exponentially:

$$S(t) = S_0 \cdot e^{rt}$$

where S_0 is the initial strength of the habit. If $S_0 = 1$ and $r = 0.05$, the strength after 30 days is:

$$S(30) = 1 \cdot e^{0.05 \times 30} \approx 4.48$$

This model suggests that consistent completion over time dramatically increases habit strength.

2.1.3   The Role of Rewards and Positive Reinforcement

Rewards serve as positive reinforcement in the habit loop, increasing the likelihood of repeating the behavior. By analyzing the reward's impact quantitatively, we can understand how reinforcement frequency impacts habit strength.

Example 1: Reward-Based Probability Model Let $R(n)$ be the probability of performing the habit after n successful completions, where each reward increases the probability by a factor k. If initially $R(0) = 0.6$ and $k = 0.1$, then:

$$R(n) = 0.6 + n \times 0.1$$

After 4 repetitions, the probability becomes:

$$R(4) = 0.6 + 4 \times 0.1 = 1.0$$

This model suggests that regular rewarding raises the likelihood of habit persistence up to a point of automaticity.

2.1.4   Tracking Habit Cues and Modifying Triggers

By identifying and modifying habit triggers, we can optimize cues to make habit initiation more reliable. Mathematical tracking of trigger effectiveness over time helps refine cues for better consistency.

Example 1: Calculating Trigger Effectiveness Suppose a trigger (e.g., an alarm) has an effectiveness rate T of 0.7. If reinforcing the cue increases effectiveness by 10

$$T(3) = 0.7 \times (1.1)^3 \approx 0.93$$

Tracking this improvement shows how small adjustments lead to habit stability over time.

2.2   The Mathematics of Habit Formation

Understanding the mathematical principles underlying habit formation can help us create and reinforce productive behaviors. Concepts like exponential growth, compounding effects, and small incremental improvements illustrate how small, consistent actions can lead to significant long-term results.

2.2.1   Exponential Growth in Habit Strength

Habits, like investments, grow over time through consistent effort. With each repetition, a habit becomes stronger, which can be modeled through exponential growth. Exponential growth suggests that the strength of a habit can increase more quickly as it compounds over time.

Example 1: Exponential Habit Growth Model

Suppose the strength of a habit is initially $S_0 = 1$, and each repetition increases the likelihood of performing the habit by a small rate r. We can model the growth of habit strength $S(t)$ after t repetitions using an exponential function:

$$S(t) = S_0 \cdot e^{rt}$$

For instance, if $r = 0.05$, the strength after 30 days (assuming the habit is performed daily) is:

$$S(30) = 1 \cdot e^{0.05 \times 30} \approx 4.48$$

This model shows how daily effort leads to rapid habit strengthening over time, as small actions compound.

2.2.2   Compounding Effects of Consistent Habits

Compounding describes how regular, repeated behaviors build upon themselves, producing increasingly significant effects over time. In habit formation, each repetition reinforces the habit loop, making future repetitions easier and more automatic.

Example 2: Compounding Effect of Habit Completion

Let's say each time you perform a habit, it increases your likelihood of performing it again by a factor of 5

$$P(n) = P(0) \cdot (1 + 0.05)^n$$

If you perform the habit daily for 10 days, the probability on day 10 becomes:

$$P(10) = 0.6 \cdot (1.05)^{10} \approx 0.98$$

This calculation illustrates that, by compounding, the likelihood of sticking with the habit approaches certainty over time.

2.2.3   The Power of Small Wins and Incremental Gains

Small, consistent improvements—referred to as "small wins"—can accumulate to create meaningful progress. In mathematical terms, small gains that add up over time yield substantial results, even if individual gains seem negligible.

Example 3: Accumulating Small Wins with Percentage Gains

Assume you improve in a certain task related to your habit by just 1

$$P(n) = 100 \cdot (1.01)^n$$

For example, after 30 days of improving by just 1

$$P(30) = 100 \cdot (1.01)^{30} \approx 134.78$$

This example shows that even minor daily improvements compound significantly, leading to major progress over time.

2.2.4   Modeling Small Habitual Improvements

While exponential models show the power of compounding, some habits grow in a linear rather than exponential manner. Linear models can be useful when the rate of improvement is steady and predictable.

Example 4: Linear Habit Improvement Model

Suppose each repetition of a habit increases your skill level by a constant amount of $d = 2$. If your initial skill level is $S_0 = 10$, then after n days, the skill level $S(n)$ follows a linear pattern:

$$S(n) = S_0 + n \cdot d$$

For example, after 15 days, your skill level becomes:

$$S(15) = 10 + 15 \cdot 2 = 40$$

This model reflects steady progress, showing that even without compounding, consistent effort leads to substantial improvements.

2.3   Practical Tools for Tracking Progress

Tracking progress in habit formation helps to reinforce consistency and provides clear feedback on improvement. Two practical tools for tracking are habit streaks, which focus on consecutive completion, and percentage-based goal achievement, which measures progress as a proportion of the intended goal.

2.3.1   Habit Streaks

A habit streak refers to the number of consecutive days or instances that a habit has been completed without interruption. Tracking streaks can serve as motivation and increase habit stability.

Example 1: Calculating Habit Streak Probability

Suppose the probability of completing a habit on any given day is $P(H) = 0.85$. The probability of maintaining a 5-day streak is the probability of completing the habit for 5 consecutive days:

$$P(\text{5-day streak}) = P(H)^5 = 0.85^5 \approx 0.4437$$

This calculation indicates that consistency decreases over a longer streak, highlighting the importance of tracking and reinforcing each daily completion.

Example 2: Increasing Streak Consistency Through Reminders

To increase streak consistency, you could use daily reminders. Assume reminders improve daily habit completion from $P(H) = 0.85$ to $P(H) = 0.9$. The probability of maintaining a 5-day streak now becomes:

$$P(\text{5-day streak}) = 0.9^5 \approx 0.5905$$

This shows that even small increases in daily success can significantly improve longer streaks.

2.3.2   Percentage-Based Goal Achievement

Percentage-based goal achievement measures how close you are to a specific target, allowing you to quantify progress in terms of completion percentage. This approach is particularly helpful when habits are not daily but require reaching a total number or amount within a set timeframe.

Example 3: Calculating Weekly Goal Achievement

Suppose your goal is to exercise 5 days per week. If you achieve 3 days, the percentage of goal achievement for the week is:

$$\text{Goal Achievement (\%)} = \frac{\text{Days Completed}}{\text{Goal Days}} \times 100 = \frac{3}{5} \times 100 = 60\%$$

This percentage shows your weekly progress and can help you adjust efforts to meet your goal fully.

Example 4: Monthly Goal Achievement for Reading Pages

Assume you have a monthly reading goal of 300 pages. If you read 240 pages by the end of the month, your goal achievement is calculated as follows:

$$\text{Goal Achievement (\%)} = \frac{240}{300} \times 100 = 80\%$$

Tracking progress as a percentage allows you to see if you are on track or need to adjust efforts to reach the target by the deadline.

2.3.3 Using Habit Streaks and Percentage-Based Tracking

Combining streaks and percentage-based goal achievement provides a well-rounded view of progress. Streaks motivate consistency, while percentage-based metrics quantify achievement relative to a defined goal.

Example 5: Weekly Tracking of a Water-Intake Goal

If your weekly goal is to drink 14 liters of water, you can use percentage tracking to measure how close you are to this target. Suppose you drink 11 liters in the week:

$$\text{Goal Achievement (\%)} = \frac{11}{14} \times 100 \approx 78.57\%$$

In addition, if you track daily streaks for completing a minimum daily water intake, you build consistency over time, reinforcing this habit while seeing weekly progress.

Chapter 3

Decision-Making and Game Theory in Daily Life

this chapter explores how to enhance decision-making through strategic thinking and game theory principles. It highlights common cognitive biases and traps, such as confirmation bias, anchoring bias, and the sunk cost fallacy, offering strategies to mitigate their effects. The chapter also delves into the application of game theory in personal and professional decisions, providing tools for anticipating others' actions and optimizing outcomes. By recognizing these biases and applying game theory, individuals can make more informed, rational decisions in various contexts.

3.1 Applying Game Theory to Make Better Decisions

Game theory, the study of strategic interactions, provides frameworks to help individuals make more informed decisions by anticipating the actions and responses of others. By understanding different game scenarios, we can apply game theory to everyday situations to enhance decision-making in competitive, cooperative, or mixed contexts.

3.1.1   Basic Concepts of Game Theory in Decision-Making

Game theory uses mathematical models to analyze situations where individuals or groups have conflicting interests, and each participant's outcome depends on the choices of others. Two commonly used types of games in decision-making are: - **Zero-sum games**: where one participant's gain is exactly balanced by the losses of others. - **Non-zero-sum games**: where participants can achieve mutually beneficial outcomes.

Example 1: The Prisoner's Dilemma and Cooperation

Consider two individuals who are suspected of a crime and are being questioned separately. They have two options: cooperate with each other by staying silent or betray each other by confessing. The payoffs for each choice are as follows:

- If both stay silent, they each receive a light sentence (1 year).

- If one betrays while the other stays silent, the betrayer is freed, and the silent party receives a heavy sentence (5 years).

- If both betray, they each receive a moderate sentence (3 years).

This scenario can be modeled as follows:

	Partner Stays Silent	Partner Betrays
You Stay Silent	$-1, -1$	$-5, 0$
You Betray	$0, -5$	$-3, -3$

In this setup, betraying is the "dominant strategy" since it minimizes risk, but mutual cooperation (both staying silent) would yield a better collective outcome. Understanding this dilemma can help us recognize situations in daily life where cooperation might be more beneficial than pursuing individual gain.

3.1.2   Applying Game Theory to Everyday Negotiations

Game theory principles are often applied to negotiation settings, where each party wants to maximize their benefits. Recognizing that negotiations are usually non-zero-sum games can help us aim for mutually beneficial outcomes.

Example 2: Negotiating with a Colleague on Task Distribution

Suppose you and a colleague need to divide tasks for a shared project. Both of you prefer certain tasks over others, but if you both pursue only your preferences, the project could suffer. By applying a "cooperative game theory" approach, you could aim for a distribution that satisfies both parties. You can: - Assign values to each task based on preference or difficulty. - Work together to allocate tasks so that both parties achieve a balanced workload. By modeling the situation as a cooperative game, you aim to maximize the overall efficiency and satisfaction rather than focusing only on individual preferences. This approach can lead to a better outcome for both participants.

3.1.3   Using Payoff Matrices to Analyze Outcomes

A payoff matrix is a tool in game theory that represents the potential outcomes for different choices, helping you visualize the consequences of each decision. This can be useful in decisions involving choices that affect both you and others.

Example 3: Choosing Between Competing and Cooperating in a Group Project: Imagine you're part of a team working on a project, and each member must decide whether to compete (focus on individual tasks and stand out) or cooperate (share resources and work collaboratively). The possible outcomes are represented by a payoff matrix:

	Team Cooperates	Team Competes
You Cooperate	$+10, +10$	$-5, +15$
You Compete	$+15, -5$	$+5, +5$

In this matrix: - If everyone cooperates, each member gets a positive outcome (+10), as collaboration benefits all. - If you cooperate while others compete, you get a negative result (-5), as others may exploit your cooperation for individual gains. - If both you and the team compete, the result is moderate (+5) but not optimal.

This matrix helps you see that cooperation yields the highest mutual benefit. By recognizing these dynamics, game theory aids in making decisions that consider the broader impact on group goals.

3.2   Strategic Thinking in Personal and Professional Contexts

Strategic thinking involves planning and decision-making that account for long-term goals, potential outcomes, and the actions of others. In both personal and professional contexts, strategic thinking allows individuals to make well-informed choices, maximize their resources, and anticipate future challenges.

3.2.1   Applying Strategic Thinking in Personal Life

In personal life, strategic thinking can be applied to achieve life goals, manage finances, and improve relationships. It involves setting clear objectives, considering alternative paths, and assessing potential risks and benefits.

Example 1: Budgeting for Long-Term Financial Goals

Suppose you want to save $10,000 over the next year. By breaking down this goal, you can create a monthly savings target of approximately $833.33. Strategic thinking would involve assessing your income, identifying spending adjustments, and creating a budget plan. By prioritizing essentials and allocating resources accordingly, you can maximize savings without sacrificing other important needs. This systematic approach helps ensure you meet the goal in a realistic and structured way.

3.2.2 Strategic Thinking in Professional Settings

In a professional context, strategic thinking is key for project planning, resource management, and long-term career growth. Effective strategy in the workplace often includes analyzing competitors, setting measurable objectives, and leveraging team strengths.

Example 2: Planning a Project Timeline with Dependencies

Imagine you are managing a project that has multiple stages, with each stage dependent on the completion of the previous one. Using strategic thinking, you can develop a project timeline by identifying key milestones and allocating resources for each stage. If Stage 1 is expected to take 2 weeks and Stage 2 requires 3 weeks, planning in advance helps you avoid delays and allocate team members efficiently to maintain progress.

3.2.3 Using Game Theory for Strategic Decision-Making

In both personal and professional settings, game theory principles can be applied to enhance strategic thinking by anticipating the actions of others. Game theory helps in scenarios where the outcome depends on the choices of multiple individuals.

Example 3: Negotiating a Salary Increase

If you are negotiating a salary, both you and your employer have specific goals. By applying a strategic approach, you can assess the employer's perspective, anticipate potential counter-offers, and decide your response thresholds. Preparing with this framework improves your position and helps achieve a favorable outcome by considering both your needs and the employer's constraints.

3.3   Common Traps and How to Avoid Them

Decision-making is often influenced by unconscious biases and cognitive traps that can lead to suboptimal choices. By recognizing these biases and applying strategies to counteract them, individuals can make more rational and effective decisions. This section explores some of the most common biases and decision-making traps, along with methods for avoiding them.

3.3.1   Confirmation Bias

Confirmation bias occurs when individuals seek out or interpret information that confirms their existing beliefs, while disregarding evidence that contradicts them. This can lead to poor decision-making by reinforcing incorrect assumptions or ignoring important factors.

Example 1: Investment Decisions

Suppose an individual is considering investing in a particular stock. They may focus only on positive news or trends related to the stock, ignoring negative reports or signs of market downturns. This bias can result in poor financial decisions. To avoid confirmation bias, it's important to seek out balanced, diverse perspectives and consider all available information before making a decision.

3.3.2   Anchoring Bias

Anchoring bias occurs when individuals rely too heavily on the first piece of information they encounter (the "anchor") when making decisions, even if that information is irrelevant or misleading. This can lead to skewed judgments and suboptimal choices.

Example 2: Pricing Decisions

Imagine you are shopping for a product and see that its original price is $200, but it is on sale for $100. The initial price ($200) acts as an anchor, making the sale price seem like a great deal, even if the actual value of the product is much lower. To counter anchoring bias, it's important to evaluate the product or decision independently, without being unduly influenced by initial information.

3.3.3 Overconfidence Bias

Overconfidence bias occurs when individuals overestimate their own abilities, knowledge, or predictions, leading to overly optimistic decisions or risky behavior. This bias can result in underestimating risks or overlooking critical details.

Example 3: Risky Financial Decisions

An individual might believe they can consistently predict stock market trends, leading them to invest heavily without proper research or diversification. Overconfidence could cause them to take on excessive risk, ultimately resulting in significant losses. To avoid overconfidence bias, individuals should seek objective feedback, gather comprehensive data, and regularly reassess their decisions.

3.3.4 Loss Aversion

Loss aversion refers to the tendency to fear losses more than valuing equivalent gains. People are generally more motivated to avoid losses than to acquire gains of the same magnitude, which can lead to overly conservative or risk-averse decisions.

Example 4: Selling Investments Prematurely

An investor may be reluctant to sell a stock at a loss, even if the stock is unlikely to recover, because the pain of realizing the loss is greater than the satisfaction of a gain. This can lead to holding onto poor investments too long. To mitigate loss aversion, it's essential to separate emotional reactions from decision-making by focusing on rational, data-driven analysis.

3.3.5 Sunk Cost Fallacy

The sunk cost fallacy occurs when individuals continue a course of action based on past investments (time, money, effort) rather than on the potential future benefits. This leads to irrational decisions because past costs cannot be recovered.

Example 5: Continuing a Failing Project

If a business has already invested significant resources into a failing project, decision-makers might continue to throw money at it simply to avoid "wasting" the initial investment, even though the future prospects are poor. To avoid the sunk cost fallacy, focus on evaluating the decision based on future benefits rather than past expenditures.

3.3.6 How to Avoid These Biases

To minimize the impact of these biases on decision-making, individuals can: - Take time to gather a wide range of information before making a decision. - Seek diverse perspectives to challenge preexisting beliefs. - Focus on objective criteria and avoid emotionally-driven decisions. - Regularly reassess past decisions and be open to changing course when new evidence arises. - Implement decision-making frameworks or checklists to ensure all relevant factors are considered.

Chapter 4

Time Management and Scheduling

This chapter introduces mathematical models for optimizing time management, including the Pareto Principle and the Eisenhower Matrix. It discusses how to prioritize tasks based on urgency and importance and provides methods like task batching, time blocking, and the Pomodoro Technique to maximize productivity. The chapter emphasizes the value of focusing on high-impact tasks and minimizing distractions, using quantitative tools to enhance decision-making and streamline daily routines for better efficiency and effectiveness in personal and professional contexts.

4.1 Mathematical Models for Optimizing Time

Effective time management requires a structured approach that helps prioritize tasks, optimize productivity, and ensure that time is spent on what truly matters. Mathematical models such as the Pareto Principle and Eisenhower Matrix provide frameworks for understanding how to allocate time more efficiently and make better decisions about task prioritization.

4.1.1 The Pareto Principle

The Pareto Principle, also known as the 80/20 rule, suggests that 80**Example 1: Applying the Pareto Principle to Work Tasks**

In a typical workday, you may have a long to-do list, but not all tasks will contribute equally to your overall success. By identifying the 20

Mathematical Expression:

If you have a list of tasks with associated time and productivity values, the Pareto Principle can be modeled as follows:

$$\sum_{i=1}^{k} T_i = 0.8 \times \sum_{i=1}^{n} T_i$$

Where T_i represents time spent on the tasks, n is the total number of tasks, and k is the subset of tasks that contribute 80

4.1.2 The Eisenhower Matrix

The Eisenhower Matrix is a time management tool that categorizes tasks into four quadrants based on their urgency and importance: - **Quadrant 1 (Urgent and Important)**: Tasks that require immediate attention. - **Quadrant 2 (Not Urgent but Important)**: Tasks that are important for long-term goals but can be scheduled. - **Quadrant 3 (Urgent but Not Important)**: Tasks that demand attention but do not significantly impact long-term objectives. - **Quadrant 4 (Not Urgent and Not Important)**: Tasks that are distractions and should be minimized.

Example 2: Using the Eisenhower Matrix for Task Prioritization

Imagine you have the following tasks: 1. Answering emails (urgent but not important)

2. Preparing for a big presentation (urgent and important)

3. Planning long-term career goals (not urgent but important)

4. Watching a TV show (not urgent and not important)

By categorizing these tasks using the Eisenhower Matrix, you can prioritize the presentation preparation (Quadrant 1), schedule the career planning (Quadrant 2), delegate or minimize the emails (Quadrant 3), and eliminate the TV show (Quadrant 4).

Mathematical Representation:

The Eisenhower Matrix can be quantified by assigning a weight to tasks based on their urgency and importance:

$$Eisenhower\ Score = \text{Urgency} \times \text{Importance}$$

Where Urgency and Importance are rated on a scale from 1 to 10. Tasks in Quadrant 1 will have the highest scores, indicating they should be prioritized.

4.1.3   Combining Both Models

Both the Pareto Principle and the Eisenhower Matrix are valuable tools for time management, and using them together can help optimize daily schedules. For example, after categorizing tasks with the Eisenhower Matrix, you can apply the Pareto Principle to focus on the most impactful tasks within Quadrants 1 and 2.

Example 3: Optimizing Daily Workflows

You may start by identifying your most important tasks (Quadrant 2) and then use the Pareto Principle to focus on the few tasks that yield the greatest return on investment in terms of effort and time. By doing so, you ensure that your time is spent on what matters most and avoid getting bogged down by distractions.

4.2   Using Productivity Ratios to Evaluate Task Value

Productivity ratios are useful tools for measuring how efficiently time and resources are being utilized in completing tasks. By calculating productivity ratios, individuals can evaluate the value and effectiveness of different tasks, allowing them to focus on high-value activities and minimize wasteful efforts. This section introduces common productivity ratios and their applications in time management.

4.2.1 Productivity Ratio: Output per Unit of Input

A basic productivity ratio can be defined as the amount of output produced per unit of input, typically time, effort, or resources. It helps evaluate the effectiveness of a task in terms of the results achieved relative to the time and resources spent.

Example 1: Calculating Productivity for a Task

Suppose you are working on writing a report. You spend 4 hours writing, and the output is a 10-page report. The productivity ratio can be calculated as:

$$\text{Productivity Ratio} = \frac{\text{Output}}{\text{Input}} = \frac{10\,\text{pages}}{4\,\text{hours}} = 2.5\,\text{pages per hour}$$

This ratio helps you assess whether you are spending an efficient amount of time relative to the output. If you want to improve your productivity, you can aim to increase the number of pages written per hour or reduce distractions that reduce your efficiency.

4.2.2 Efficiency Ratio

The efficiency ratio compares the time spent on a task relative to the value it provides. A high efficiency ratio indicates that you are getting significant output for relatively little time, whereas a low ratio suggests inefficiencies that should be addressed.

Example 2: Evaluating Efficiency in Task Completion

Imagine you are managing several tasks: one involves preparing a presentation, which takes 5 hours, and another involves responding to routine emails, which takes 2 hours. The perceived value of the presentation is much higher than the emails. You might calculate an efficiency ratio based on time spent and value delivered, assigning values to the tasks:

$$\text{Efficiency Ratio} = \frac{\text{Value of Task}}{\text{Time Spent}}$$

For the presentation, if you assign a value of 50 (out of 100) and it takes 5 hours, the efficiency ratio is:

$$\text{Efficiency Ratio (Presentation)} = \frac{50}{5} = 10$$

For the emails, if you assign a value of 10 and it takes 2 hours, the efficiency ratio is:

$$\text{Efficiency Ratio (Emails)} = \frac{10}{2} = 5$$

This shows that the presentation is more efficient in terms of value per time spent, and therefore should be prioritized over responding to emails.

4.2.3   Return on Time Invested (ROTI)

Return on Time Invested (ROTI) is a specific productivity ratio that evaluates the return on time spent on a task, often used to assess the value or profit gained from time invested. It can be particularly useful in decision-making, where you compare different tasks to maximize value.

Example 3: Calculating ROTI for a Project

Imagine you are working on a project that costs $200 in time (e.g., 10 hours at $20/hour) and yields $1000 in value. The ROTI is calculated as:

$$\text{ROTI} = \frac{\text{Value}}{\text{Time Invested}} = \frac{1000}{200} = 5$$

This means that for every hour of time invested, the return is 5 times the value, indicating a high return on time.

4.2.4   Task Value Ratio

The Task Value Ratio is used to determine the relative value of different tasks. By calculating the task value ratio, individuals can determine which tasks provide the highest return in terms of the time spent on them.

Example 4: Calculating Task Value Ratio

You are juggling multiple projects. The first project requires 10 hours to complete and has a value of 40 units. The second project requires 5 hours to complete and has a value of 30 units. The Task Value Ratios are:

$$\text{Task Value Ratio (Project 1)} = \frac{40}{10} = 4 \, \text{units per hour}$$

$$\text{Task Value Ratio (Project 2)} = \frac{30}{5} = 6 \, \text{units per hour}$$

In this case, Project 2 provides a higher value per hour spent, suggesting that it may be more beneficial to focus on Project 2 in order to maximize productivity.

4.2.5   Improving Productivity Using Ratios

By calculating and tracking these ratios regularly, individuals can identify areas where improvements can be made. For example, if a particular task has a low productivity or efficiency ratio, it might be an indication that the task is not being executed in the most efficient way, or it might not be worth investing the same amount of time. Optimizing workflows, eliminating unnecessary steps, and leveraging technology can all help increase productivity ratios over time.

4.3   Methods for Prioritizing and Batching Tasks

Efficient task management involves not only prioritizing tasks but also grouping similar tasks together (batching) to reduce cognitive load and increase productivity. In this section, we explore different methods and strategies for prioritizing and batching tasks, using both quantitative and qualitative approaches to maximize output and minimize wasted time.

4.3.1   The ABCDE Method for Task Prioritization

The ABCDE method is a simple, yet effective, way to prioritize tasks based on their importance and urgency. Tasks are assigned one of five labels:
- A: Must-do tasks with serious consequences if not completed.
- B: Should-do tasks with mild consequences if not completed.
- C: Nice-to-do tasks with no serious consequences.
- D: Delegate tasks that can be handled by someone else.
- E: Eliminate tasks that are unnecessary or do not contribute to your goals.

Example 1: Using the ABCDE Method

You have the following tasks for the day: 1. Finish an important project proposal (A)

2. Answer non-urgent emails (B)

3. Organize your workspace (C)

4. Respond to a request for a meeting that can be handled by someone else (D)

5. Watch a TV show (E)

By labeling tasks according to their importance and urgency, you can quickly identify what should be prioritized. In this case, focus on the project proposal (A), delegate the meeting request (D), and eliminate the TV show (E).

Mathematical Representation:

You can assign numerical values to tasks to quantify their priority:

$$\text{Priority Score} = \text{Urgency} \times \text{Importance}$$

Tasks with the highest scores are prioritized.

4.3.2   The Eisenhower Matrix

The Eisenhower Matrix is another effective tool for prioritizing tasks. It divides tasks into four quadrants based on their urgency and importance: - Quadrant 1 (Urgent and Important): Tasks that need immediate attention.

- Quadrant 2 (Not Urgent but Important): Tasks that contribute to long-term goals. - Quadrant 3 (Urgent but Not Important): Tasks that require attention but do not contribute significantly to long-term goals.

- Quadrant 4 (Not Urgent and Not Important): Tasks that are distractions or can be eliminated.

Example 2: Using the Eisenhower Matrix for Prioritization

Tasks for the day might include: 1. Responding to an urgent client email (Urgent Important)

2. Writing a long-term business strategy (Not Urgent Important)

3. Answering low-priority emails (Urgent Not Important)

4. Checking social media (Not Urgent Not Important)

By categorizing these tasks into quadrants, you can prioritize Quadrant 1 tasks immediately, schedule Quadrant 2 tasks, delegate Quadrant 3 tasks, and eliminate Quadrant 4 tasks.

4.3.3   Task Batching

Batching involves grouping similar tasks together and performing them consecutively without interruption. This minimizes the time spent switching between different types of tasks, allowing you to focus on similar tasks that require the same cognitive processes.

Example 3: Batching Email Responses

Rather than checking emails throughout the day, set aside a specific block of time (e.g., 30 minutes) to batch all email responses. This allows you to handle multiple emails in one go, saving time and mental energy compared to constantly switching between tasks.

Mathematical Insight:

Batching tasks can reduce the total time spent on tasks by reducing the cost of task-switching, often modeled as:

$$\text{Total Time} = \sum (\text{Time per Task}) - \text{Time Lost to Switching}$$

By batching, the time lost to switching tasks is minimized, resulting in a lower total time for completing tasks.

4.3.4 Time Blocking

Time blocking is a method where you allocate specific blocks of time in your schedule for specific tasks or groups of tasks. This method ensures that you focus on one task at a time and helps prevent multitasking, which can often lead to inefficiency.

Example 4: Using Time Blocking for Task Management

You might schedule: - 9:00 AM - 11:00 AM: Focus on writing a report.

- 11:00 AM - 12:00 PM: Respond to emails (batching).

- 12:00 PM - 1:00 PM: Lunch.

- 1:00 PM - 3:00 PM: Complete client calls.

By dedicating uninterrupted time to each type of task, you can work more efficiently without the distractions of switching between different tasks.

4.3.5 The Pomodoro Technique

The Pomodoro Technique involves working in focused intervals (typically 25 minutes), followed by short breaks (5 minutes). After completing four Pomodoros, a longer break (15-30 minutes) is taken. This method encourages sustained focus and helps prevent burnout.

Example 5: Using the Pomodoro Technique for Task Completion

If you're working on a report, you might set a timer for 25 minutes of focused work. After 25 minutes, you take a 5-minute break. After completing four Pomodoros, you take a longer break. This structured approach helps maintain high productivity levels over long periods.

4.3.6 Task Prioritization and Batching Strategies

By combining task prioritization and batching methods, individuals can achieve optimal productivity. First, use a method like the ABCDE method or Eisenhower Matrix to prioritize tasks based on their importance and urgency. Then, batch similar tasks together and allocate dedicated time blocks for their completion. This approach minimizes distractions, reduces task-switching, and ensures that high-priority tasks are given adequate focus.

Chapter 5

The Mathematics of Goal Setting and Tracking

This chapter focuses on how mathematical principles can aid in setting, breaking down, and tracking goals. By applying concepts such as SMART goals, progressions, ratios, and feedback loops, this chapter teaches how to measure progress and adjust goals dynamically. It covers the use of mathematical models to structure large goals into manageable steps, ensuring continual motivation and systematic achievement. The chapter integrates both quantitative and qualitative feedback for effective goal management.

5.1   SMART Goals and the Math Behind Effective Goal Setting

Setting clear, achievable, and measurable goals is crucial for success in both personal and professional life. The SMART framework is widely used to define effective goals by ensuring they are Specific, Measurable, Achievable, Relevant, and Time-bound. This section explores the mathematics behind setting SMART goals and how you can apply these principles to maximize productivity and success.

5.1.1 The SMART Framework

SMART goals are built on five key criteria: - **Specific**: The goal must be clearly defined and unambiguous. - **Measurable**: The goal must be quantifiable to track progress. - **Achievable**: The goal must be realistic and attainable. - **Relevant**: The goal must align with long-term objectives. - **Time-bound**: The goal must have a deadline for completion.

Example 1: Setting a SMART Goal for Fitness

Suppose your goal is to improve your fitness. A SMART goal might look like:

Goal: Increase my running distance to 5 kilometers in 3 months, running 3 times a week.

Breaking this down: - **Specific**: Increase running distance to 5 km. - **Measurable**: Track the distance run each session. - **Achievable**: Gradually increase the distance by 0.5 km per week. - **Relevant**: The goal aligns with overall health and fitness objectives. - **Time-bound**: Complete in 3 months.

5.1.2 The Mathematics of Measurability

One of the key elements of SMART goals is measurability. Quantifying progress helps track how close you are to achieving the goal, and it ensures accountability. Mathematical methods, such as tracking increments, ratios, and percentages, are often used to monitor progress over time.

Example 2: Quantifying Progress in Learning a New Skill

If your goal is to learn a new language within 6 months, you might set measurable milestones:

Goal: Learn 1000 new words in 6 months.

To track progress, break the goal down into smaller chunks:

$$\text{Words per week} = \frac{1000}{6 \times 4} = 41.67 \text{ words per week}$$

After each week, check how many words you've learned and compare it to the target of 41.67 words.

5.1.3 Time-Bound Goals: Deadlines and Time Allocation

A crucial part of SMART goals is that they are time-bound, meaning they have specific deadlines. Time-bound goals help create a sense of urgency and prevent procrastination. By allocating time to different phases of a goal, you can ensure steady progress.

Example 3: Time Allocation for Project Completion

If you are working on a project with a deadline of 4 weeks, and the total task involves 100 hours of work, you can calculate the time to spend each week:

$$\text{Hours per week} = \frac{100 \text{ hours}}{4 \text{ weeks}} = 25 \text{ hours per week}$$

By allocating 25 hours per week, you ensure that you stay on track to complete the project on time.

5.1.4 The Math of Achievability

Achievability ensures that goals are not too far out of reach. It involves considering your current abilities, available resources, and constraints. By using incremental milestones, you can break down larger goals into smaller, more manageable parts.

Example 4: Achievable Milestones for a Sales Target

Suppose your goal is to increase sales revenue by 20% in a year. Break this down into quarterly targets:

$$\text{Quarterly Target} = \frac{20\% \text{ increase}}{4 \text{ quarters}} = 5\% \text{ increase per quarter}$$

By setting achievable milestones of 5% increase every quarter, you make the overall goal more attainable.

5.1.5 Tracking Progress with Ratios and Percentages

Tracking progress involves using ratios, percentages, and other mathematical tools to evaluate how much of a goal has been achieved. These tools provide a clear, quantifiable way to assess performance over time.

Example 5: Using Ratios to Track Financial Goals

If you set a financial goal to save $5000 in 6 months, you can track your savings progress:

$$\text{Savings per month} = \frac{5000}{6} = 833.33 \text{ dollars per month}$$

After each month, compare your actual savings with the target, and calculate the percentage of the goal achieved:

$$\text{Percentage of Goal Achieved} = \frac{\text{Actual Savings}}{\text{Target Savings}} \times 100$$

If after 3 months you have saved $2500, then:

$$\text{Percentage of Goal Achieved} = \frac{2500}{5000} \times 100 = 50\%$$

This helps you evaluate your progress and make adjustments as needed.

5.1.6 Goal Tracking Software and Mathematical Tools

In addition to manual tracking, there are various goal-setting software tools that use mathematical models to help you visualize progress. These tools typically display data in charts, graphs, and progress bars to give you a clearer picture of your achievements over time.

5.2 Breaking Large Goals into Achievable Steps

Large goals can often feel overwhelming, but breaking them down into smaller, more manageable steps makes them more achievable. By using progressions, milestones, and ratios, you can create a structured plan for achieving your larger goals. This approach not only makes progress more visible but also keeps you motivated as you reach each milestone along the way.

5.2.1 Progressions

Progressions involve breaking a large goal into smaller, incremental steps. This method ensures that you are continuously moving toward your goal, even when it may feel distant. A common approach is to use arithmetic or geometric progressions to set measurable milestones.

Example 1: Using an Arithmetic Progression to Achieve a Fitness Goal

Suppose your goal is to run 100 kilometers over the next 10 weeks. Using an arithmetic progression, you could increase the distance by 10 kilometers each week:

$$\text{Distance per week:} \quad 10, 20, 30, 40, 50, 60, 70, 80, 90, 100$$

This breakdown ensures that each week's goal is a manageable and measurable increase. By the end of week 10, you will have achieved the total goal of 100 kilometers.

Mathematical Insight:

If the goal is broken into n steps, where the first step is a and the common difference between steps is d, the total goal can be calculated by the sum of

the progression:
$$S_n = \frac{n}{2}(2a + (n-1)d)$$

In this case, $a = 10$, $d = 10$, and $n = 10$, so the total sum (goal) is:

$$S_{10} = \frac{10}{2}(2(10) + (10-1)10) = 550 \text{ kilometers}$$

5.2.2 Milestones

Milestones are specific targets within the larger goal that help track progress. By setting clear milestones, you can measure how much progress you have made and adjust your strategy as necessary. These milestones act as checkpoints, keeping you motivated and ensuring that the larger goal is on track.

Example 2: Setting Milestones for a Business Revenue Target

Suppose your business goal is to generate $50,000 in revenue over the next year. Setting quarterly milestones will allow you to track progress:

$$\text{Quarter 1:} \quad \$12,500, \text{Quarter 2:} \quad \$12,500$$

$$\text{Quarter 3:} \quad \$12,500, \text{Quarter 4:} \quad \$12,500$$

By breaking down the larger goal into quarterly milestones, you ensure that each quarter is a step toward your final target, and you can assess whether adjustments need to be made in terms of marketing, sales strategies, or other aspects.

5.2.3 Ratios: Tracking Progress Using Percentages

Ratios are a simple and effective way to track progress toward a goal by calculating the percentage of the goal completed at any given point. By continuously evaluating progress, you can determine how close you are to achieving your objective and make adjustments if necessary.

Example 3: Using Ratios to Track Progress

Suppose your goal is to save $5,000 over six months. At the end of each month, you calculate how much money you've saved and use the ratio to determine your progress:

$$\text{Progress Ratio} = \frac{\text{Amount Saved}}{\text{Goal}} \times 100$$

For example, if you've saved $2,500 after three months:

$$\text{Progress Ratio} = \frac{2500}{5000} \times 100 = 50\%$$

This percentage tells you that you are halfway to achieving your goal. If you are ahead or behind schedule, you can adjust your savings plan for the remainder of the time.

5.2.4 Cumulative Tracking: Progress Over Time

Cumulative tracking involves calculating the total progress over time by adding up the achievements of each smaller step. This method ensures that you see the total amount of work done and provides a visual representation of progress toward the larger goal.

Example 4: Cumulative Progress in a Learning Goal

If your goal is to read 12 books in a year, you could track your progress cumulatively by reading one book per month. After six months, you would have read 6 books, meaning you have completed 50% of the goal:

$$\text{Cumulative Progress} = \frac{\text{Books Read}}{\text{Total Books}} \times 100$$

$$\text{Cumulative Progress} = \frac{6}{12} \times 100 = 50\%$$

This method allows you to assess your progress as a whole while keeping track of how much you've achieved over time.

5.2.5   Using Mathematical Models for Goal Breakdown

Mathematical models, such as linear programming or optimization techniques, can be applied to break down large goals into smaller achievable steps. These models help allocate resources, manage time, and balance multiple goals effectively, ensuring that each part of the larger goal is achievable with the available resources.

Example 5: Optimization of Resource Allocation for Multiple Goals
Suppose you are working toward several goals simultaneously, such as completing a project, learning a new skill, and exercising regularly. You can use linear programming to allocate your available time efficiently across these tasks, ensuring that each goal receives the attention needed for optimal progress.

5.3   Methods for Measuring Progress

Measuring progress toward a goal is essential for staying on track, and adapting your goals based on feedback is a key element in achieving success. Feedback loops help you assess your performance, identify areas for improvement, and make adjustments as necessary to ensure that your goals remain realistic and achievable. This section explores different methods for measuring progress and adapting your goals based on feedback.

5.3.1   Tracking Progress with Quantitative Metrics

One of the most effective methods for measuring progress is by using quantitative metrics. These can include numerical targets, percentages, or any other measurable quantities that allow you to track how much of the goal has been achieved. Regularly evaluating these metrics ensures that you can determine if you're on track or need to adjust your approach.

Example 1: Tracking Sales Progress

Suppose your goal is to increase monthly sales by 20%. You can track progress by comparing actual sales against the target:

$$\text{Sales Progress} = \frac{\text{Actual Sales}}{\text{Target Sales}} \times 100$$

For example, if the target sales for the month are $10,000 and the actual sales are $8,000:

$$\text{Sales Progress} = \frac{8000}{10000} \times 100 = 80\%$$

This progress percentage tells you that you are at 80% of your goal, indicating the need to adjust sales strategies or focus on higher-value prospects to reach the 100% target.

5.3.2   Setting Milestones and Feedback Points

Setting specific milestones along the way allows you to assess your progress periodically and adapt your strategies accordingly. Feedback points are pre-set moments where you pause to evaluate how well you are moving toward your goal. These points can trigger adjustments if you find that progress is slower or faster than expected.

Example 2: Milestones in a Learning Goal

If your goal is to complete a certification course in 6 months, you could set milestones to assess your progress every month:

Month 1: Complete Module 1, Month 2: Complete Module 2

Month 3: Complete Module 3

After each milestone, evaluate how much you've learned and whether you are on schedule. If progress is slower than expected, you can adjust the pace, dedicate more time, or seek help to ensure that you stay on track.

5.3.3 Using Feedback Loops

A feedback loop involves regularly gathering data on your progress and using this information to make informed adjustments. The idea is to assess your current performance, analyze the results, and modify your approach to improve future performance. This cycle of feedback and adjustment is crucial for continuous improvement.

Example 3: Feedback Loop in Fitness Training

Suppose your goal is to increase your strength in weightlifting by 15% over 3 months. You can measure progress by tracking the weights lifted every week. After each week, you evaluate whether you're making sufficient progress. If you find that you're not progressing fast enough, you may adjust your training routine by adding variety, changing the intensity, or increasing rest periods.

Mathematically, you can model your progress using a growth function, such as:

$$\text{Progress}_t = \text{Initial Weight} \times (1 + r)^t$$

where r is the rate of progress per week, and t is the time in weeks. After every feedback cycle, you can adjust r to reflect your actual rate of improvement and make necessary changes.

5.3.4 Dynamic Goal Adjustment Based on Performance

As you progress toward your goal, it is important to be flexible and adjust your goals based on performance. If you are achieving your milestones faster than expected, you can raise your target; conversely, if you're falling behind, you can reassess and modify the goal to make it more achievable. This adaptive approach helps maintain motivation and ensures that your goals remain realistic and challenging.

Example 4: Adjusting Financial Goals:

If your financial goal is to save $6,000 in 6 months and you are able to save $1,200 in the first month (20% ahead of target), you can adjust the goal for the remaining months:

$$\text{Remaining Target} = 6000 - 1200 = 4800$$

If you continue to save at this faster rate, you can increase your target savings for the remaining months. However, if you fall behind, you can lower the target to make it more achievable.

5.3.5 Incorporating Statistical Models to Predict Progress

Statistical models, such as regression analysis or moving averages, can be used to predict progress and adapt goals based on past performance. These models can help identify trends and outliers in your progress and offer insights into how much progress can realistically be made over time.

Example 5: Predicting Progress in a Project

If you're working on a long-term project with several phases, you can use regression analysis to track how much work is being completed over time. For instance, if you have data points showing progress every week, a linear regression model can be used to predict future progress:

$$y = mx + b$$

where y is the predicted progress, m is the slope (rate of change), and b is the y-intercept (initial progress). By adjusting the rate of change m based on feedback, you can predict and modify your approach to ensure timely completion of the project.

5.3.6 Adapting to Feedback

Feedback can be both quantitative (measurable data) and qualitative (subjective insights). Both types of feedback are valuable in adapting your approach and improving the chances of success. Quantitative feedback allows for adjustments based on clear data, while qualitative feedback often involves intuition, feelings of satisfaction, or other non-numeric indicators.

Example 6: Combining Quantitative and Qualitative Feedback

Suppose you're tracking how much time you spend on tasks each day. Quantitative feedback might show that you're spending more time on low-priority tasks than expected. Qualitative feedback could reveal that you're feeling overwhelmed and unmotivated. By using both types of feedback, you can adjust your schedule to spend more time on high-priority tasks and reduce distractions.

Chapter 6

Budgeting and Financial Planning for Personal Growth

Thischapter explores using financial metrics to assess spending and align finances with personal values. It introduces key metrics like savings rate, debt-to-income ratio, and spending ratios to evaluate how well your spending habits support your goals. The chapter demonstrates how to model monthly expenses, track progress toward savings, and create a value-based financial plan. By regularly reviewing and adjusting your financial decisions, you can ensure your spending aligns with long-term priorities, such as health, family, or financial independence, helping you make more intentional financial choices.

6.1 Budgeting Basics and the Power of Compounding

Budgeting is a key component of financial planning that allows you to allocate resources effectively, track your expenses, and ensure that you are moving toward your financial goals. Understanding the basics of budgeting, along with the power of compounding, can help you maximize your financial growth and achieve long-term personal success.

6.1.1   The Basics of Budgeting

Budgeting is the process of creating a plan for managing your income and expenses. It helps ensure that your spending aligns with your financial goals. A basic budget divides your income into categories such as savings, living expenses, and discretionary spending. This division helps you prioritize your needs and ensure financial stability.

Example 1: Simple Budgeting Strategy

Suppose you have a monthly income of $3,000. You could allocate the following:

$$\text{Savings:}\quad 20\% \quad \text{of Income}\quad = 600$$

$$\text{Living Expenses:}\quad 50\% \quad \text{of Income}\quad = 1500$$

$$\text{Discretionary Spending:}\quad 30\% \quad \text{of Income}\quad = 900$$

This budget ensures that 20% of your income is saved, which is crucial for long-term financial growth.

6.1.2   The Role of Compounding in Personal Finance

Compounding refers to the process where the value of an investment grows exponentially over time due to the earning of interest on both the initial principal and the accumulated interest from previous periods. This concept can be applied to savings, investments, and even debt repayment.

Example 2: Power of Compounding in Savings

Suppose you invest $1,000 in a savings account with an annual interest rate of 5%, compounded annually. After one year, the interest earned would be:

$$\text{Interest} = 1000 \times 0.05 = 50$$

At the end of the year, your balance will be:

$$\text{Balance} = 1000 + 50 = 1050$$

In the second year, the interest will be calculated on the new balance of $1,050, leading to further exponential growth. Over time, compounding allows your savings to grow faster than simple interest, significantly increasing the amount accumulated.

6.1.3 Compounding in Investments

Investing money in assets such as stocks, bonds, or real estate, and allowing the returns to compound over time can result in substantial wealth accumulation. The longer the time horizon, the more significant the impact of compounding.

Example 3: Compounding in Investment

Consider investing $5,000 at an annual return rate of 8%, compounded yearly. After 10 years, the value of the investment would be:

$$\text{Future Value} = 5000 \times \left(1 + \frac{8}{100}\right)^{10} = 5000 \times (1.08)^{10} = 10,794.62$$

The effect of compounding is apparent as the value of the investment nearly doubles in 10 years, showing how time and consistent returns can lead to wealth accumulation.

6.1.4 Budgeting for Growth and Financial Independence

In addition to saving and investing, effective budgeting involves planning for long-term financial growth and independence. By setting aside funds for retirement, emergency savings, and major life goals, you create a pathway for financial stability. Compound interest can play a significant role in building wealth for the future.

Example 4: Retirement Planning with Compounding

If you contribute $200 per month to a retirement account that earns an annual return of 7%, compounded monthly, after 30 years you will have:

$$\text{Future Value} = 200 \times \frac{\left(1 + \frac{0.07}{12}\right)^{12 \times 30} - 1}{\frac{0.07}{12}} = 200 \times 938.22 = 187,644$$

By consistently contributing and allowing the power of compounding to work over time, you can accumulate substantial wealth for retirement.

6.1.5 Debt Management and the Impact

While compound interest can be a powerful tool for wealth building, it can also work against you when applied to debts. High-interest loans and credit cards often compound interest, increasing the amount you owe exponentially if not paid off quickly.

Example 5: Debt Accumulation with Compound Interest

Suppose you have a credit card balance of $2,000 with an annual interest rate of 18%, compounded monthly. After one year, the debt would grow as follows:

$$\text{Debt after 1 Year} = 2000 \times \left(1 + \frac{18}{100 \times 12}\right)^{12}$$

$$= 2000 \times (1.015)^{12} = 2,361.21$$

The total debt increases due to the effect of compound interest. Therefore, timely repayment is crucial to avoid exponential growth of debt.

6.2 Modeling Monthly Expenses, Savings Goals

Modeling monthly expenses, setting savings goals, and financial forecasting are crucial steps in effective financial planning. By using mathematical models and financial formulas, you can better predict your future financial state, track your expenses, and plan for long-term growth.

6.2.1   Modeling Monthly Expenses

The first step in financial planning is understanding where your money goes each month. Monthly expenses can be categorized into fixed costs (e.g., rent, utilities) and variable costs (e.g., groceries, entertainment). By modeling these expenses, you can identify areas where you can reduce costs and allocate more money toward savings and investments.

Example 1: Modeling Monthly Expenses

Suppose you have the following monthly expenses:

$$\text{Fixed Expenses:Rent: } 1200, \text{ Utilities: } 150, \text{ Insurance: } 100$$

$$\text{Variable Expenses:Groceries: } 300, \text{ Transportation: } 100, \text{ Entertainment: } 200$$

Total monthly expenses can be modeled as:

$$\text{Total Monthly Expenses} = \text{Fixed Expenses} + \text{Variable Expenses}$$

$$\text{Total Monthly Expenses} = (1200 + 150 + 100) + (300 + 100 + 200) = 2050$$

This model shows you are spending \$2,050 each month, which is crucial for adjusting your savings goals.

6.2.2   Setting Savings Goals and Tracking Progress

To set effective savings goals, it's important to understand your monthly income and expenses. A good starting point is to aim to save a percentage of your income each month. A common guideline is to save 20% of your income.

Example 2: Setting and Tracking Savings Goals

If your monthly income is \$3,000 and you aim to save 20%, the savings goal is:

$$\text{Savings Goal} = 3000 \times 0.20 = 600$$

This means that you should aim to save $600 each month. After modeling your income and expenses, you can track whether you are meeting your savings goal. If your expenses are higher than anticipated, you may need to adjust your discretionary spending to maintain this savings rate.

6.2.3   Financial Forecasting for Long-Term Goals

Financial forecasting allows you to predict your future financial state based on current data, such as savings rates, investment returns, and expenses. This can help you plan for major life events, like buying a home, retirement, or funding education.

To model long-term savings, we can use the formula for the future value of an investment with regular contributions, including compound interest:

$$\text{Future Value} = P \times \left(1 + \frac{r}{n}\right)^{nt} + PMT \times \frac{\left(1 + \frac{r}{n}\right)^{nt} - 1}{\frac{r}{n}}$$

where: - P is the initial investment - r is the annual interest rate (decimal) - n is the number of times interest is compounded per year - t is the number of years - PMT is the regular monthly contribution

Example 3: Financial Forecasting for Retirement: Suppose you invest $5,000 initially with monthly contributions of $200, an annual interest rate of 5%, and the investment compounds monthly for 30 years. Using the formula, the future value of the investment would be:

$$\text{Future Value} = 5000 \times \left(1 + \frac{0.05}{12}\right)^{12 \times 30} + 200 \times \frac{\left(1 + \frac{0.05}{12}\right)^{12 \times 30} - 1}{\frac{0.05}{12}}$$

$$\text{Future Value} = 5000 \times (1.004167)^{360} + 200 \times \frac{(1.004167)^{360} - 1}{0.004167}$$

$$\text{Future Value} \approx 5000 \times 4.467 + 200 \times 716.39 = 22,335 + 143,278 = 165,613$$

This model predicts that after 30 years of saving $200 per month at an annual interest rate of 5%, your total investment will grow to $165,613, illustrating the power of compounding.

6.2.4 Modeling Debt Repayment and Interest Impact

In addition to savings, financial planning involves managing debt. Modeling debt repayment and understanding how compound interest can affect debt accumulation is crucial to avoid financial setbacks. When making payments on loans or credit cards, it is important to model the impact of interest over time.

Example 4: Debt Repayment with Interest: If you have a $10,000 credit card balance with an interest rate of 18% per year, compounded monthly, the debt will grow as follows if no payments are made:

$$\text{Debt after 1 year} = 10000 \times \left(1 + \frac{0.18}{12}\right)^{12} = 10000 \times (1.015)^{12} = 11,933.06$$

This shows that after one year, your balance will increase to $11,933.06 due to the compounded interest, making timely payments crucial for managing debt.

6.2.5 Tracking and Adjusting Your Financial Plan

It is important to track your financial plan regularly and adjust your income, expenses, and savings goals based on actual performance. This allows you to stay on track and make necessary changes to ensure financial success.

Example 5: Tracking Progress Toward Savings Goal

If your monthly savings goal is $600, but you only saved $500 in the first month, you can make adjustments by saving $700 in the next month to stay on track. This ensures that you meet your annual savings target despite fluctuations in monthly savings.

6.3 Using Metrics to Assess Spending

Assessing your spending habits and aligning them with your personal values is a critical part of financial planning. By using metrics and analytical tools, you can ensure that your spending reflects your priorities and helps you achieve your long-term goals. This section explores how to use metrics to assess spending patterns and align them with your values, leading to better financial decision-making.

6.3.1 Understanding Financial Alignment

Aligning your finances with your values means spending money in ways that support your personal goals and beliefs. For example, if personal growth, family, and health are important values, your spending should reflect these priorities. Financial metrics help quantify how well your spending habits align with your values, giving you a clearer picture of where adjustments may be needed.

Example 1: Aligning Spending with Personal Values

Let's say that your core values are health, personal development, and family. You might prioritize spending on a gym membership, books, and family vacations. By comparing your actual spending to your values, you can identify if too much money is being spent on discretionary items (e.g., dining out, entertainment) and make adjustments to align more closely with your values.

6.3.2 Metrics for Assessing Spending Patterns

To evaluate spending, various financial metrics can be used. These metrics provide insights into where money is going and how well it aligns with your goals. Common metrics include:

Savings Rate: The percentage of income saved each month.

Debt-to-Income Ratio: A measure of how much of your income is spent on servicing debt.

Spending Ratio: The proportion of income allocated to essential versus non-essential expenses.

Example 2: Calculating Spending Ratios

Suppose you have a monthly income of $3,000, and you spend:

$$\text{Essential Expenses: } 2000, \quad \text{Non-Essential Expenses: } 800$$

Your spending ratio for essential expenses is:

$$\text{Essential Spending Ratio} = \frac{\text{Essential Expenses}}{\text{Total Income}} = \frac{2000}{3000} = 0.67 \ (67\%)$$

Your spending ratio for non-essential expenses is:

$$\text{Non-Essential Spending Ratio} = \frac{\text{Non-Essential Expenses}}{\text{Total Income}}$$

$$= \frac{800}{3000} = 0.27 \ (27\%)$$

By using these ratios, you can see if your spending aligns with your values. If family and health are top priorities, you may choose to reallocate some of the non-essential spending towards health-related or family-related activities.

6.3.3   Assessing Financial Goals Through Metrics

Using metrics to assess your financial goals helps ensure that you are on track to meet them. For instance, if saving for a house, measuring your progress through metrics such as the **Savings Rate** and the **Amount Saved vs. Goal** will give you a clear indication of whether your current savings strategies align with your objective.

Example 3: Progress Towards Savings Goal

Suppose your goal is to save $10,000 for a down payment on a house. After six months, you have saved $5,000. Your savings progress can be calculated as:

$$\text{Progress Percentage} = \frac{\text{Amount Saved}}{\text{Goal Amount}} \times 100 = \frac{5000}{10000} \times 100 = 50\%$$

This metric shows you are halfway to your goal and allows you to reassess your spending and savings rate if you're falling behind.

6.3.4 Creating a Financial Plan that Reflects Your Values

Once you have assessed your spending using these metrics, it's important to create a financial plan that reflects your values. This plan should prioritize your top values—whether it's savings, investment, health, or family—and allocate resources accordingly. Financial planning should be a dynamic process, adjusting to changes in your goals and values over time.

Example 4: Creating a Value-Based Financial Plan

If you value financial independence and security, your financial plan might allocate:

50% for Savings and Investments, 30% for Essential Expenses

20% for Non-Essential Expenses (e.g., entertainment, luxury items)

By applying this plan, you are ensuring that the majority of your resources are allocated to long-term financial security, with enough flexibility for discretionary spending that still aligns with your values.

6.3.5   Adjusting Spending Based on Financial Metrics

After tracking and assessing your spending habits, adjustments may be necessary to ensure your spending is aligned with your financial goals and values. For example, if you find that you are spending too much on non-essential items, you can choose to cut back and reallocate that money toward savings or investments.

Example 5: Adjusting Spending for Better Alignment

If your analysis shows that you spend 40% of your income on non-essential items but would prefer to allocate more toward savings, you could aim to reduce discretionary spending to 20%. This change would increase your savings rate from 10% to 30%, helping you reach your financial goals more quickly.

Chapter 7

Quantifying Health and Fitness Habits

This chapter focuses on quantifying health and fitness habits by applying mathematical metrics like Body Mass Index (BMI) and resting heart rate. It explains how these metrics are calculated and interpreted, providing context with statistical averages. By tracking these health indicators, individuals can set realistic goals, monitor progress, and make informed decisions to improve their overall well-being. The chapter emphasizes the importance of data collection and offers practical examples for using health metrics to guide fitness routines and enhance health outcomes.

7.1 Applying Math to Fitness Goals

In this chapter, we explore how mathematics can be used to enhance health and fitness habits by quantifying key metrics such as calories burned, steps taken, and performance progress. By understanding the mathematical principles behind these metrics, you can make more informed decisions about your fitness routines and track your progress toward achieving your health goals.

7.1.1 Calories Burned: A Mathematical Approach

Understanding the calories burned during various activities is crucial for achieving fitness goals related to weight loss, weight maintenance, or muscle gain. The number of calories burned depends on factors like exercise intensity, duration, and individual characteristics such as weight and age. The following formula can be used to estimate the calories burned during exercise:

$$\text{Calories Burned} = \text{MET} \times \text{Weight (kg)} \times \text{Duration (hours)}$$

Where: - MET (Metabolic Equivalent of Task) represents the intensity of the activity. - Weight is in kilograms. - Duration is the time spent performing the activity in hours.

Example 1: Calculating Calories Burned During Running

Suppose you weigh 70 kg and run for 45 minutes at a pace that corresponds to a MET value of 9 (moderate running). The number of calories burned would be:

$$\text{Calories Burned} = 9 \times 70 \times 0.75 = 472.5 \text{ calories}$$

This calculation helps quantify the energy expenditure from your activity.

7.1.2 Steps Taken

Tracking steps is a simple yet effective way to measure daily physical activity. The goal is often to reach 10,000 steps per day, but this can vary depending on individual fitness goals. Tracking steps can be used to monitor progress and ensure you are staying active throughout the day.

Example 2: Estimating Calories Burned from Walking

To estimate the calories burned from walking, you can use a general formula:

$$\text{Calories Burned from Walking} = \text{Step Count} \times \text{Calories per Step}$$

Assuming you burn approximately 0.04 calories per step, if you walk 10,000 steps, the calories burned would be:

$$\text{Calories Burned} = 10,000 \times 0.04 = 400 \text{ calories}$$

This simple calculation helps you understand how much energy you expend throughout the day and how it contributes to your overall fitness goals.

7.1.3   Performance Tracking

Performance tracking goes beyond just calories and steps; it involves measuring improvements in strength, endurance, flexibility, and other physical capacities. Key metrics to track include: - Repetitions and Sets: For strength training, tracking the number of sets and reps allows you to measure progress. - Distance and Time: For endurance activities like running, cycling, or swimming, tracking distance covered and time taken allows you to assess improvements in performance.

Example 3: Tracking Strength Progress

Suppose you're doing squats and want to track your progress. In your first workout, you perform 3 sets of 10 repetitions with a weight of 50 kg. In your next workout, you do 3 sets of 10 repetitions with 60 kg. The progression can be quantified as:

$$\text{Progress} = \frac{\text{New Weight}}{\text{Old Weight}} = \frac{60}{50} = 1.2 \text{ (20\% increase in weight)}$$

This metric helps you track strength gains over time.

7.1.4   Optimizing Fitness Plans

Mathematical models can also help optimize your fitness plans by considering various factors like intensity, rest periods, and frequency of workouts. For example, the **FIT** formula, which stands for **Frequency, Intensity, and Time**, can be used to structure a balanced exercise routine.

$$FIT = Frequency \times Intensity \times Time$$

Where: -frequency is the number of exercise sessions per week. -Intensity is how hard the workout feels, often measured in heart rate zones or MET values. -Time is the duration of each session.

Example 4: Optimizing Exercise Routine

Suppose you plan to exercise 4 days a week, with an intensity of 70% of your maximum heart rate, for 45 minutes each session. The FIT model would look like:

$$FIT = 4 \times 0.7 \times 45 = 126$$

This formula gives you a numerical value for the overall intensity of your fitness program, helping you assess if it's balanced and effective for achieving your fitness goals.

7.1.5   Using Technology for Tracking and Analysis

With modern technology, tracking and analyzing health and fitness metrics has become easier. Wearables like fitness trackers and smartwatches can automatically track steps, heart rate, calories burned, and even sleep patterns. Many apps also provide detailed insights into your workouts, allowing you to adjust your routine and track your progress over time.

7.2   The Importance of Data Collection for Health Goals

Data collection is a cornerstone of effective health and fitness planning. By systematically tracking key metrics, individuals can gain valuable insights into their progress, identify areas for improvement, and make data-driven decisions to achieve their health goals. This section explores the importance of data collection and how it can help optimize your fitness journey.

7.2.1 Tracking Key Metrics

For health and fitness, key metrics often include calories burned, steps taken, heart rate, sleep patterns, and exercise performance. By collecting data on these metrics, you can monitor your progress and make informed adjustments to your routines.

Example 1: Tracking Calories Burned

If you are trying to lose weight, tracking calories burned through exercise and daily activities helps you ensure you're in a calorie deficit. For example, by tracking calories burned during exercise and comparing it to your daily caloric intake, you can determine if you are on track to achieve weight loss.

7.2.2 Identifying Patterns and Trends

Data collection allows you to identify patterns and trends in your health and fitness habits. For example, you might notice that you perform better in the gym after getting a full night's sleep or that your running speed improves after consistently walking 10,000 steps per day.

Example 2: Identifying Performance Trends

By tracking your running times over several weeks, you might notice that your pace improves as you consistently complete 3 runs per week. This pattern can provide motivation and help you adjust your training routine to maintain progress.

7.2.3 Making Data-Driven Adjustments

Having data on your health habits allows you to make informed decisions and adjust your plans to stay on track with your goals. Whether you're trying to increase your physical activity, improve your diet, or enhance your sleep, tracking data helps you understand what's working and what isn't.

Example 3: Adjusting Workout Intensity

Suppose your goal is to increase endurance. By tracking your performance in cardio workouts, such as the distance or time spent running, you can assess whether your intensity level needs to be increased to continue progressing. If your times have plateaued, you can adjust by adding more challenging workouts or varying the intensity.

7.2.4 Using Technology for Efficient Data Collection

Advancements in technology have made data collection more accessible and efficient. Fitness trackers, wearable, and health apps can automatically track key metrics such as steps, calories burned, and heart rate, providing a wealth of information at your fingertips. These tools help you keep a detailed record of your activities and progress, making it easier to assess and adjust your routines.

Example 4: Using Wearable for Health Data

Using a fitness tracker, you can easily collect data on your daily step count, heart rate, calories burned, and even sleep patterns. For example, a Fit bit or Apple Watch will provide real-time feedback and historical data that can be used to assess how well you are meeting your health goals, allowing for quick adjustments when necessary.

7.2.5 Establishing Baselines

One of the first steps in data collection is establishing a baseline. By collecting data over time, you can create a clear picture of your starting point and set realistic, measurable goals. Regularly revisiting this data helps you assess whether you are meeting your targets or if changes need to be made.

Example 5: Establishing a Baseline for Exercise

If you want to increase your strength, start by tracking your current performance, such as the amount of weight lifted in specific exercises. After a few weeks of training, compare your progress to the baseline to evaluate how much improvement you've made. This comparison allows you to set new, realistic goals and further refine your fitness routine.

7.3 Practical Examples

Health metrics are valuable tools for quantifying and assessing your physical well-being. In this section, we will explore practical examples of commonly used health metrics such as **Body Mass Index (BMI)** and **resting heart rate**, and how statistical averages can be used to interpret these metrics. These metrics provide a way to evaluate your fitness level, track progress, and make informed decisions regarding your health.

7.3.1 Body Mass Index (BMI)

The Body Mass Index (BMI) is a simple yet effective metric used to assess an individual's weight relative to their height. While it does not directly measure body fat, it provides a useful approximation for categorizing individuals into various weight categories such as underweight, normal weight, overweight, and obese.

The formula for calculating BMI is:

$$BMI = \frac{\text{Weight (kg)}}{(\text{Height (m)})^2}$$

Example 1: Calculating BMI

Suppose you weigh 70 kg and are 1.75 meters tall. Your BMI can be calculated as follows:

$$BMI = \frac{70}{(1.75)^2} = \frac{70}{3.0625} \approx 22.86$$

According to BMI categories, this is classified as **Normal weight** (18.5–24.9).

7.3.2 Resting Heart Rate (RHR)

Resting heart rate (RHR) is a key indicator of cardiovascular fitness. It measures the number of heartbeats per minute when you are at rest. A lower resting heart rate generally indicates better cardiovascular fitness, as the heart is able to pump blood more efficiently. The normal range for adults is typically between 60 and 100 beats per minute (bpm), but athletes may have a resting heart rate as low as 40-60 bpm.

Example 2: Measuring Resting Heart Rate

To measure your resting heart rate, sit quietly for at least 5 minutes and then count your pulse for 60 seconds. If your pulse is 72 beats per minute, your resting heart rate is 72 bpm, which falls within the normal range for most adults.

7.3.3   Statistical Averages

When analyzing health metrics, statistical averages can provide context and benchmarks for understanding your results. For example, average BMI values and resting heart rate ranges can help you assess where you stand compared to the general population.

-Average BMI: The average BMI for adults globally is around **24.9**, which places most individuals in the "Normal weight" category. -Average Resting Heart Rate: The average resting heart rate for adults is typically **72 bpm**, with variations based on age, fitness level, and gender.

Example 3: Interpreting Health Metrics with Statistical Averages

If your BMI is 28, it is classified as **Overweight**. Since the average BMI is 24.9, your BMI is above the statistical average, which may indicate an increased risk for certain health conditions, such as heart disease or diabetes. Similarly, if your resting heart rate is 85 bpm, it is above the average resting heart rate of 72 bpm, suggesting that your cardiovascular fitness could be improved.

7.3.4   Using Averages to Set Health Goals

Understanding the statistical averages for various health metrics helps you set realistic goals. For example, if you are aiming to improve your cardiovascular fitness, you might set a goal to reduce your resting heart rate over time through consistent aerobic exercise.

Example 4: Setting a Resting Heart Rate Goal

If your current resting heart rate is 80 bpm, a realistic goal might be to reduce it by 5-10 bpm over the next 3 months. You can achieve this by engaging in regular cardio workouts, such as running, cycling, or swimming. Tracking your progress can be done by measuring your resting heart rate weekly and comparing it to your starting value.

Chapter 8

Measuring and Analyzing Productivity with Analytics

This chapter explores the use of analytics and statistical tools to measure and analyze productivity. It focuses on three key visualization tools: histograms, trend lines, and scatter plots, explaining how each can be used to track productivity patterns, recognize trends, and identify inefficiencies. The chapter emphasizes the importance of data-driven decisions for improving productivity by visually representing data and understanding the relationships between variables. By leveraging these tools, readers can gain insights into their work habits and optimize their performance for greater efficiency.

8.1   Using Statistics to Track and Improve Productivity

Statistics provide a powerful way to measure and analyze productivity. By applying statistical concepts such as mean, median, and mode, you can gain insights into your work patterns, identify trends, and improve efficiency. This section explores how to use these fundamental statistics to track and optimize productivity.

8.1.1   Mean: The Average Productivity Level

The mean, or average, is one of the most commonly used statistics in productivity analysis. It provides an overall measure of your productivity by calculating the sum of your outputs over a specific period and dividing it by the number of periods. The formula for calculating the mean is:

$$\text{Mean} = \frac{\sum \text{Productivity Outputs}}{\text{Number of Periods}}$$

Example 1: Calculating the Mean Productivity

Suppose you tracked the number of tasks completed over five workdays: 8, 10, 12, 7, and 9 tasks. The mean productivity can be calculated as follows:

$$\text{Mean} = \frac{8 + 10 + 12 + 7 + 9}{5} = \frac{46}{5} = 9.2$$

This means your average productivity is 9.2 tasks per day.

8.1.2   Median: The Middle Point of Productivity Data

The median is the middle value in a set of numbers when they are arranged in ascending or descending order. It is particularly useful when analyzing productivity data with outliers, as it is less affected by extreme values compared to the mean.

Example 2: Calculating the Median Productivity

Using the same productivity data: 8, 10, 12, 7, and 9 tasks. First, arrange the numbers in ascending order: 7, 8, 9, 10, 12. The median is the middle value, which is **9** tasks. This shows that 50

8.1.3   Mode: The Most Frequent Productivity Level

The mode represents the most frequent value in a dataset. In the context of productivity, the mode can highlight the most common level of output or the most frequent type of task completed.

Example 3: Calculating the Mode of Productivity

Consider the following data on tasks completed over a week: 8, 8, 12, 7, 9, 8, 10. The mode of this dataset is **8**, as it appears most frequently. This indicates that on most days, you complete around 8 tasks.

8.1.4   Using Statistics to Improve Productivity

By understanding and analyzing the mean, median, and mode of your productivity data, you can make informed decisions to enhance your work habits. For instance, if your productivity is consistently below your desired level (mean), you can identify areas where improvement is needed. Conversely, if the mode is much higher than the mean, it might indicate that a certain type of task is being completed more often than others, and adjustments can be made to balance your workload.

Example 4: Improving Productivity with Data Insights

If your goal is to increase daily productivity to 10 tasks, but your average (mean) is only 8 tasks, you can analyze your workflow to identify factors that are slowing you down. Is it time management, distractions, or inefficient processes? By identifying patterns in the data, you can take action to optimize your workflow and achieve your desired productivity levels.

8.1.5   Practical Tools for Productivity Analytics

Several tools, including spreadsheets, time-tracking software, and task management apps, can help collect and analyze productivity data. These tools often have built-in functions to calculate mean, median, and mode, allowing you to easily track your progress and identify areas for improvement.

Example 5: Using Productivity Software

Apps like Trello, To doist, orescue Time can track your tasks and time usage, automatically calculating your average productivity. By analyzing this data over time, you can identify trends and make adjustments to improve your efficiency.

8.2   Understanding Your Productivity Trends

Understanding your productivity trends and patterns is essential for optimizing your workflow and maximizing efficiency. By analyzing your
productivity data over time, you can identify recurring behaviors, peak
performance times, and areas that may need improvement. This section
explains how to track and interpret productivity patterns to make informed
decisions that enhance your overall performance.

8.2.1   Identifying Peak Performance Times

One of the most valuable insights you can gain from tracking your productivity is recognizing the times of day when you are most productive.
Productivity is often cyclical, and identifying peak periods can help you
allocate more challenging or high-priority tasks during those times.

Example 1: Analyzing Peak Productivity Periods

Suppose you track your productivity (measured by tasks completed) over
a week and find that on average, you complete: - 4 tasks in the morning
(8 AM – 12 PM), - 6 tasks in the afternoon (12 PM – 4 PM), - 2 tasks in the
evening (4 PM – 8 PM).

This trend indicates that your peak productivity period is in the afternoon.
Recognizing this, you can schedule your most demanding tasks during this
time to optimize your efficiency.

8.2.2   Tracking Task Types and Their Impact

By analyzing the types of tasks you complete and the time spent on each, you
can identify patterns in how different task categories impact your overall
productivity. For example, routine tasks might be quicker to complete
but less productive, while complex tasks might take longer but yield more
significant results.

Example 2: Analyzing Task Categories

Suppose you categorize your tasks as follows: - Routine tasks: 5 tasks (3 hours) - High-priority tasks: 3 tasks (5 hours) - Administrative tasks: 4 tasks (2 hours)

By tracking these categories over time, you might notice that while high-priority tasks take longer, they significantly contribute to your overall productivity and long-term goals. This insight can help you prioritize high-priority tasks over routine ones for maximum impact.

8.2.3   Recognizing Trends in Efficiency

By plotting productivity data over time, you can identify trends such as improving or declining efficiency. You might find that your productivity has been steadily increasing, or you could notice periods of stagnation or decline. These trends can help you understand whether your productivity is improving with new strategies or if adjustments are needed.

Example 3: Analyzing Productivity Trends

If you track the number of tasks completed daily over a month and notice that your productivity spikes at the beginning of the month but declines after two weeks, it could indicate burnout or a lack of motivation. Recognizing this pattern allows you to take proactive steps, such as scheduling breaks or altering your workload distribution to maintain consistent productivity.

8.2.4   Using Data to Identify Areas for Improvement

Understanding your productivity patterns can reveal areas for improvement. If you consistently find that certain types of tasks or time periods are less productive, you can experiment with new strategies, tools, or schedules to address these inefficiencies.

Example 4: Identifying Bottlenecks in Productivity

Suppose you notice that you consistently complete fewer tasks on Mondays than on other days of the week. This might indicate that the start of the week is difficult for you due to low energy or motivation. By recognizing this pattern, you can adjust your Monday schedule to focus on simpler tasks that require less effort, allowing you to ease into the workweek and gradually increase your productivity.

8.2.5   Visualizing Productivity Trends with Graphs

Visualizing productivity data through graphs and charts can provide a clearer understanding of trends and patterns. Tools like spreadsheets or productivity apps can generate graphs that help you track your productivity over time, making it easier to spot fluctuations, trends, and areas for improvement.

Example 5: Visualizing Trends

Using a spreadsheet, you can create a line graph showing your daily productivity (e.g., number of tasks completed). The graph might reveal that your productivity tends to dip at certain points during the day or week. By visualizing these trends, you can take actionable steps to address these fluctuations, such as adjusting your work schedule or introducing new time management techniques.

8.3   Tools for Visualizing Productivity

Visualization is a powerful tool for understanding and improving productivity. By graphically representing productivity data, you can easily identify trends, patterns, and anomalies that are difficult to notice in raw data alone. In this section, we will explore three common visualization tools—**histograms**, **trend lines**, and **scatter plots**—and how they can be used to analyze and enhance productivity.

8.3.1 Histograms

A histogram is a type of bar chart that represents the frequency distribution of a set of data. In the context of productivity, histograms can help you visualize how often certain productivity levels (e.g., number of tasks completed per day) occur. This tool is useful for identifying the most common productivity levels and spotting potential outliers or trends.

Example 1: Creating a Histogram of Daily Task Completion

Suppose you tracked the number of tasks completed over 30 days, and the data is as follows: [5, 6, 7, 8, 8, 9, 10, 11, 12, 5, 7, 9, 6, 10, 8, 11, 12, 7, 9, 6, 5, 7, 10, 9, 8, 7, 11, 6, 9, 5] A histogram of this data might show a concentration of tasks completed between 7 and 10, with fewer days spent on extreme values like 5 or 12 tasks. This insight allows you to understand how often you complete specific numbers of tasks and can help identify if your productivity is consistent or fluctuating.

8.3.2 Trend Lines: Visualizing Productivity Over Time

Trend lines are graphical representations of data trends over time. By plotting productivity data on a graph and fitting a trend line, you can easily see whether your productivity is increasing, decreasing, or remaining stable. Trend lines are particularly useful for tracking long-term progress toward productivity goals.

Example 2: Plotting a Trend Line for Monthly Productivity

If you track the number of tasks completed each month over a year, the data might look like this: Month 1: 80, Month 2: 85, Month 3: 90, Month 4: 95, Month 5: 100, Month 6: 105, Month 7: 110, Month 8: 115, Month 9: 120, Month 10: 125, Month 11: 130, Month 12: 135

Plotting this data on a graph with a trend line will show an upward slope, indicating that your productivity is steadily increasing each month. The trend line helps you see whether you are on track to meet your yearly productivity goals and provides a visual cue for areas needing improvement.

8.3.3 Scatter Plots

Scatter plots are used to examine the relationship between two variables. By plotting data points on a two-dimensional graph, you can identify correlations or patterns between different factors influencing productivity. For example, you could use a scatter plot to explore how time spent on a task correlates with the quality of the output or how different working conditions affect productivity.

Example 3: Using a Scatter Plot to Examine Time Spent vs. Task Completion

Suppose you track the amount of time spent on tasks and the number of tasks completed. Your data might look like this:

$$[(1, 4), (2, 7), (3, 10), (4, 11), (5, 12), (6, 14), (7, 15), (8, 16), (9, 17)]$$

In this case, the first value in each pair represents the time spent (in hours), and the second value represents the number of tasks completed. By plotting this data on a scatter plot, you can visually assess the relationship between time and task completion. A positive correlation would be evident if the points form an upward-sloping line, suggesting that as more time is spent, the number of tasks completed increases.

8.3.4 Using Tools for Visualizing Productivity Data

There are several tools and software applications that can help you create and analyze these visualizations: - Excel/Google Sheets: Both platforms offer built-in tools for creating histograms, trend lines, and scatter plots. Simply input your data, select the appropriate chart type, and customize the visualization.

Tableau: A more advanced tool that allows you to create interactive and dynamic visualizations. It can connect to multiple data sources and create complex, multi-dimensional views of productivity data.

R/Python (with Matplotlib/Seaborn): If you are familiar with programming, R and Python provide powerful libraries for statistical analysis and data visualization, including the creation of histograms, trend lines, and scatter plots.

Example 4: Using Excel for Visualization

To create a histogram in Excel: 1. Input your productivity data in a column. 2. Select the data, go to the "Insert" tab, and choose "Histogram" from the chart options. 3. Excel will automatically generate the histogram and allow you to adjust the number of bins and other visual elements.

Chapter 9

Resilience and Failure: Statistical Models of Success

this chapter focuses on Resilience and Failure through the lens of statistical models of success. It highlights how understanding and managing setbacks can foster personal growth. The chapter emphasizes the importance of building a growth mindset, where challenges and failures are seen as opportunities for learning. By using data and mathematical models such as probability, feedback loops, and exponential growth, individuals can quantify setbacks, identify patterns, and predict future challenges. It also illustrates how incremental improvements, when tracked through numbers, lead to sustained success. The chapter provides practical examples and mathematical insights to develop resilience and turn failure into a stepping stone for long-term achievement.

9.1 Understanding Probability and Failure Rates

Failure is an inevitable part of personal growth, but how you interpret and react to failure can significantly impact your success. Statistical models, particularly in probability, can help us understand the likelihood of failure and success, and how resilience plays a crucial role in overcoming setbacks.

9.1.1   The Role of Probability in Personal Growth

Probability is the measure of the likelihood that an event will occur. In personal growth, understanding probability can help you assess risks, plan for setbacks, and make decisions based on likely outcomes. By quantifying the chance of success or failure, you can prepare better strategies for achieving your goals and overcoming obstacles.

Example 1: Calculating the Probability of Success in a New Habit
Imagine you are trying to establish a new habit, such as exercising daily. Based on your past attempts, you estimate that you succeed in maintaining a new habit 60

9.1.2   Failure Rates and Resilience

Failure rates represent the frequency with which you fail to achieve a desired outcome. In personal growth, failure is not a sign of weakness but an opportunity to learn and improve. Resilience—the ability to bounce back after failure—is essential for personal development. Statistical models of failure rates can help you understand how often setbacks are likely to occur and how to adjust your expectations and strategies.

Example 2: Modeling Failure in Habit Formation
Consider that research shows 40

9.1.3   Success and Failure as Complementary Outcomes

While failure is a common outcome, success and failure are interconnected. Each failure provides valuable information about what works and what doesn't. Statistically, failure can provide a learning experience that increases the probability of success in future attempts. This is especially true in fields like business, personal development, and habit formation, where persistence and iteration are key.

Example 3: Success as the Result of Iteration

If you are working on a project and face multiple failures, each failure can be viewed as an opportunity to adjust your approach. Statistically, the more iterations you make (i.e., trying different strategies), the higher the likelihood of eventual success. For example, Thomas Edison famously failed thousands of times before inventing the lightbulb, but each failure helped refine his approach and bring him closer to success.

9.1.4   Using Probability to Manage Risk

Risk is inherent in all areas of personal growth, from financial investments to career changes or lifestyle shifts. By understanding the probability of success and failure in these areas, you can better manage risk and make informed decisions. Statistical models allow you to assess potential outcomes and minimize negative impacts from failure, helping you stay on track even when challenges arise.

Example 4: Assessing the Risk of Career Change

If you're considering a career change, you may use probability to assess the likelihood of success based on factors such as your skills, experience, and market conditions. For example, you might estimate a 70

9.1.5   Learning from Failure: The Resilience Curve

The resilience curve represents how individuals respond to failure over time. Initially, failure may lead to frustration and decreased motivation, but with practice, individuals can learn to recover more quickly and become more resilient. Statistically, the more exposure to failure you have, the better you become at managing it and learning from it. Over time, your resilience improves, making you more likely to achieve success in future endeavors.

Example 5: Building Resilience Through Repeated Attempts

Suppose you face several rejections while applying for jobs. Initially, each rejection may feel discouraging, but over time, you begin to develop a stronger sense of resilience. Statistically, the more times you face rejection and continue to try, the higher your chances of success in the future. This gradual improvement in resilience reflects a positive shift in your personal growth journey.

9.2   Modeling Setbacks and Using Data to Build Resilience

Setbacks are an inevitable part of personal growth and achievement. Understanding how to model and quantify setbacks using data can provide valuable insights into building resilience. By tracking setbacks, analyzing their causes, and adjusting your strategies, you can develop a more resilient approach to personal challenges. This section discusses how to model setbacks, use data to understand their impact, and apply statistical techniques to foster resilience and improve your chances of success in future endeavors.

9.2.1   Understanding Setbacks

Setbacks often feel like obstacles that delay progress, but they can be viewed as opportunities for learning and improvement. By quantifying setbacks, you can measure their impact and identify patterns or common causes. Statistical models allow you to analyze the frequency, severity, and outcomes of setbacks, providing insights into how they affect your overall growth.

Example 1: Quantifying Setbacks in Habit Formation

Suppose you are trying to build a new habit, such as exercising regularly, but experience setbacks (e.g., missing a day of exercise). If you track each setback, you might find that you miss an average of 1.5 days per week. Using this data, you can model the setbacks and calculate the impact on your overall progress. For instance, if you miss 1.5 days per week on average, you can estimate how many workouts you'll miss over a month and adjust your expectations accordingly.

9.2.2   Using Data to Identify the Causes of Setbacks

Analyzing data can help you identify the root causes of setbacks, allowing you to address them directly. For example, if you're consistently missing workouts, tracking data related to time, energy levels, or external distractions may reveal patterns that contribute to these setbacks. By understanding the factors that lead to setbacks, you can make more informed decisions and adopt strategies to prevent or mitigate them in the future.

Example 2: Analyzing Causes of Setbacks in Goal Achievement

Let's say you're tracking progress toward a goal, and setbacks occur regularly. By tracking factors like time of day, stress levels, or competing responsibilities, you might find that you experience setbacks during high-stress periods. This data allows you to build resilience by adjusting your strategy—for example, by planning more breaks during stressful times or adjusting your goal timeline to accommodate these challenges.

9.2.3   Statistical Models for Predicting Future Setbacks

Once you have gathered data on setbacks, you can use statistical models to predict their likelihood in the future. For example, you might use a **probability distribution** to estimate the chances of experiencing setbacks based on past data. By forecasting setbacks, you can prepare better and design strategies to minimize their impact, such as adjusting deadlines or setting buffer time for unexpected delays.

Example 3: Using Probability Distributions to Predict Setbacks

Assume you've experienced 10 setbacks over 6 months. By using a **Poisson distribution**, you can predict the likelihood of experiencing setbacks in the upcoming month. If the average rate of setbacks is 1.67 per month, the Poisson distribution can help you estimate the chances of having 0, 1, 2, or more setbacks, giving you valuable information to manage expectations and take preventive action.

9.2.4   Building Resilience

Resilience can be enhanced by using data in feedback loops. By continuously tracking progress and setbacks, you create a feedback system that helps you adapt and improve over time. When setbacks occur, data can guide you in making adjustments to your strategy, helping you recover quickly and keep moving toward your goals. This feedback loop ensures that setbacks do not derail progress but instead contribute to the learning and refinement of your approach.

Example 4:

Using Feedback Loops for Continuous Improvement

If you're tracking your productivity, and you notice a setback (e.g., a day when productivity drops significantly), you can analyze the data and make adjustments. For instance, you might find that you're working longer hours without breaks, leading to burnout. By integrating short breaks into your schedule, you can prevent future productivity setbacks. This continuous cycle of tracking, analyzing, and adjusting allows you to build resilience and improve performance over time.

9.2.5   Statistical Resilience Curve

The resilience curve demonstrates how individuals typically respond to setbacks over time. Initially, setbacks may cause frustration or negative emotions, but as individuals accumulate more experiences with failure, they develop better coping strategies and become more resilient. Statistically, the more exposure to setbacks you have, the better you become at managing and learning from them. The resilience curve suggests that failure, when viewed as a learning opportunity, can lead to greater long-term success.

Example 5: Resilience Curve in Job Search

When job hunting, you may face repeated rejections. Initially, this may feel discouraging, but over time, you begin to develop resilience. By analyzing your responses to setbacks, you realize that rejection is not personal but a part of the process. The resilience curve shows that with each rejection, your confidence and ability to cope with failure improve, leading to greater success in future job applications.

9.3   The Role of Growth Mindsets

A growth mindset, a concept popularized by psychologist Carol Dweck, refers to the belief that abilities and intelligence can be developed with effort, learning, and persistence. This mindset plays a crucial role in personal growth, particularly when setbacks occur. Viewing failures as opportunities to learn and improve is essential for long-term success. In this section, we will explore how growth mindsets and incremental improvements, when quantified through numbers, can help enhance resilience and increase the likelihood of achieving personal goals.

9.3.1   Growth Mindsets

A growth mindset encourages individuals to view challenges as opportunities to improve rather than insurmountable obstacles. When setbacks happen, individuals with a growth mindset are more likely to persevere, viewing failure as a natural part of the learning process. The role of numbers here is crucial—by quantifying progress, individuals can track their improvement and gain the confidence needed to continue pursuing their goals.

Example 1: Tracking Progress with Incremental Gains

Suppose you're learning a new skill, such as playing a musical instrument, and you initially struggle with basic chords. By tracking your practice sessions and measuring small improvements (e.g., time spent practicing, accuracy of notes played), you can visualize your progress. Even small gains, quantified over time, help reinforce the growth mindset, as you can see how incremental improvements lead to long-term success.

9.3.2   Incremental Improvement

Incremental improvement emphasizes making small, continuous progress towards a goal. Rather than attempting a major leap or perfection in a short period, incremental improvement focuses on steady, sustainable gains. This concept is closely related to the principle of **marginal gains**, which suggests that small improvements in various areas can compound over time and lead to significant overall progress. The role of numbers here is essential, as each small improvement can be tracked and quantified, providing evidence of progress.

Example 2: The Power of Small Wins in Fitness

Imagine you want to improve your running speed. Instead of trying to drastically improve your time in one run, you aim for small, incremental gains. For example, in each run, you aim to reduce your time by 10 seconds. Over the course of 10 runs, this 10-second improvement compounds, leading to a significant overall improvement in performance. By tracking each incremental gain, you can see how small changes add up over time, reinforcing the power of incremental improvement.

9.3.3 Mathematical Models

Mathematical models can quantify incremental improvements and help individuals track progress. **Exponential growth** and **compound interest** are mathematical concepts that highlight how small, consistent gains can accumulate over time. These models can be applied to personal growth in areas such as skill development, fitness, and productivity.

Example 3: Exponential Growth in Learning a New Language

Suppose you want to learn a new language and set a goal of learning 5 new words every day. If you consistently add 5 words to your vocabulary each day, the total number of words learned will grow exponentially over time. After 30 days, you will have learned 150 new words. The growth rate accelerates as time progresses.

9.3.4 Feedback Loops and Growth Mindset

The growth mindset thrives on continuous feedback. When you measure and track progress, data creates a feedback loop that informs your next steps. As you see the results of your efforts, whether positive or negative, you adjust your strategies to optimize future outcomes. This feedback loop reinforces the growth mindset by providing tangible evidence that effort leads to improvement.

Example 4: Measuring Progress in Skill Acquisition

Imagine you're learning to cook and you track various metrics such as time spent cooking, number of recipes attempted, or improvements in taste scores from family and friends. As you see small improvements in these metrics, you are encouraged to continue practicing and refining your cooking skills. Over time, your skills improve incrementally, reinforcing the belief that continued effort leads to growth.

9.3.5   The Math Behind Growth Mindset

By quantifying progress, individuals with a growth mindset can see their improvements more clearly, which boosts motivation and confidence. Numbers provide an objective measure of progress, helping to combat self-doubt. Whether it's tracking daily progress in habit formation, skill development, or fitness, seeing numerical improvements helps solidify the belief that growth is achievable with persistence.

Example 5: Confidence in Habit Formation

If you are building a habit, such as drinking 8 glasses of water every day, tracking your daily progress can reinforce the growth mindset. After a week, you may see that you've successfully completed the habit 6 out of 7 days. This numerical evidence demonstrates that you are on the right track, which boosts your confidence and encourages continued effort, even if you miss a day or two.

Chapter 10

The Feedback Loop: Calculating and Adapting Progress

This chapter focuses on how to effectively measure progress, receive feedback, and make necessary adjustments to ensure continuous improvement. It delves into the concept of the feedback loop, emphasizing the role of consistent monitoring and data analysis in personal development. Through the use of statistical techniques such as significance testing and feedback loops, individuals can determine when changes in their habits or strategies are needed. The chapter also explores how to balance adaptation with consistency to maintain sustained growth over time, offering practical mathematical models and examples to guide this process. Ultimately, the chapter aims to equip readers with the tools and insights needed to optimize their habits and achieve long-term personal success.

10.1 Measuring Feedback and Adjusting Actions

The feedback loop is a crucial concept in personal growth and productivity, as it helps individuals assess their progress and adjust their actions to achieve their goals. In this section, we will explore how to measure feedback, analyze progress, and adapt strategies to stay on track towards personal and professional success. By incorporating mathematical techniques, such as optimization, data analysis, and statistical modeling, we can fine-tune our actions and make data-driven decisions to improve performance and outcomes.

10.1.1   Understanding the Feedback Loop

A feedback loop refers to a system where outputs of an action or process are used as inputs for future actions. In the context of personal growth, feedback is the information that results from an action (e.g., completing a task or attempting a goal), which is then used to adjust and improve future actions. By incorporating feedback, we can continuously refine our approach, improving over time. The key is to measure feedback consistently and use it to inform decision-making.

Example 1: Fitness Goal Adjustment Based on Feedback

Let's say you're trying to increase your daily step count to improve fitness. Initially, you aim for 10,000 steps a day, but after tracking your progress, you find that you are consistently averaging 8,000 steps. By adjusting your goal to a more achievable target of 8,500 steps, you use the feedback from your previous progress to refine and optimize your strategy for success.

10.1.2   Using Metrics to Track Progress

To ensure that progress is measurable and actionable, it's important to use appropriate metrics. These metrics provide the necessary data to evaluate where adjustments are needed. Common metrics include time spent on tasks, completion rates, success rates, or any relevant key performance indicators (KPIs) that align with your goals. Measuring these metrics allows you to identify areas for improvement and adapt your strategies accordingly.

Example 2: Tracking Progress in Habit Formation with Metrics

If you're trying to build a habit of reading, tracking the number of pages read each day is a helpful metric. By analyzing this data, you may notice that you read more on weekends than weekdays.

10.1.3   Adapting Actions Based on Feedback

Once feedback has been collected and analyzed, the next step is to adapt your actions based on the insights gained. This may involve adjusting your goals, changing your strategy, or altering your approach to suit your progress. Adapting actions allows you to optimize your efforts and make more effective decisions that will lead to better outcomes.

Example 3: Adjusting a Study Plan Based on Feedback

Imagine you're studying for an exam and initially set a goal to study for 3 hours each day. After a week, you assess your performance and find that you're only completing 2 hours per day, and your retention rate is low. The feedback from your initial plan indicates that you may need to adjust your study method or the time spent studying. You could increase the study time, break it into smaller sessions, or incorporate more active learning techniques.

10.1.4   Optimization Techniques for Feedback Loops

In some cases, optimization techniques can be used to improve the effectiveness of feedback loops. For example, mathematical models such as **gradient descent** or **linear programming** can help optimize the process by minimizing the difference between the desired and actual outcomes. These techniques are often used in systems that require continuous improvement, like business processes, habit formation, or personal productivity.

Example 4: Optimizing Task Management with Linear Programming

If you're managing multiple tasks with different deadlines and priorities, you can use linear programming to optimize your schedule. By considering the time required for each task and the deadline constraints, you can adjust your actions in real-time to maximize productivity while meeting your goals.

10.1.5   Feedback Loops in Personal Development

Personal development relies heavily on feedback loops, as they allow individuals to assess their growth and adapt accordingly. Whether you're working on improving a skill, managing time more effectively, or pursuing a fitness goal, measuring and analyzing feedback enables continuous improvement. The use of data, mathematics, and optimization helps fine-tune actions and strategies, ensuring that progress remains on track.

Example 5:Improvement in Learning a New Skill

If you are learning a new language, tracking metrics such as vocabulary retention, grammar usage, and listening comprehension over time gives you feedback on your progress. By analyzing this feedback, you can adjust your study techniques or time allocation to ensure continuous improvement. For instance, if you find that listening comprehension is lagging behind, you can adjust your study focus to emphasize listening exercises.

10.2   Using Statistical Significance

Statistical significance plays a vital role in analyzing feedback for habit adjustments. By applying statistical principles, we can determine whether observed changes in habits are meaningful or simply due to chance. This helps in making informed decisions about when to adjust a habit or when the current strategy is working. The combination of statistical analysis and feedback provides a rigorous method to assess progress, evaluate patterns, and refine personal development strategies effectively.

10.2.1   In Habit Formation

Statistical significance refers to the likelihood that an observed effect is real and not due to random fluctuations. When tracking habits, it's important to determine whether the progress you're seeing is substantial enough to warrant an adjustment or if it's just normal variation. Using statistical tests such as **t-tests** or **confidence intervals**, you can assess whether changes in your behavior are meaningful and whether adjusting your approach is necessary.

Example 1: Evaluating Habit Formation Progress with Statistical Tests:

If you're tracking your daily exercise habits over a month, you might see fluctuations in the number of hours spent exercising each week. By calculating the average hours exercised each week and performing a t-test, you can determine whether the observed changes in exercise duration are statistically significant or if they could just be random variations.

10.2.2   Applying Feedback to Habit Adjustments

Once feedback is collected, the next step is to analyze it statistically to make informed decisions about habit adjustments. Feedback can be used to identify patterns or trends that may not be immediately obvious, such as times of the day when a habit is most successfully performed or external factors that may affect habit completion.

Example 2: Identifying Patterns in Habit Formation

Imagine you're trying to establish a reading habit and track the number of pages you read each day. After analyzing the data, you may find that you read more consistently on certain days, such as weekends, and less consistently during weekdays. By performing a statistical analysis, you can identify this pattern and adjust your habit strategy, perhaps by setting a goal to read at a specific time each day or adjusting the type of reading material.

10.2.3   Using Feedback Loops

Statistical models, such as **regression analysis** or **moving averages**, can help to predict future habit trends based on past performance. These models allow for continuous monitoring of habit progress and can highlight when a habit is deviating from expectations, indicating when adjustments are needed. By incorporating statistical significance into feedback loops, individuals can ensure that habit changes are based on data-driven insights rather than assumptions.

Example 3: Using Moving Averages to Track Habit Progress

To smooth out fluctuations in daily data and gain clearer insights, you might use a moving average to track the number of hours exercised each week. This technique helps to minimize the impact of daily variations and allows for better decision-making when making adjustments to your exercise habits. If your moving average starts to decline, it could indicate that an adjustment to your routine is needed.

10.2.4   Making Data-Driven Habit Adjustments

By combining statistical significance with feedback, individuals can make data-driven decisions to improve or adjust their habits. This approach ensures that changes are based on objective evidence rather than subjective assumptions. Whether you're aiming to improve productivity, fitness, or time management, using statistics to evaluate habit adjustments allows for more effective and efficient personal growth.

Example 4: Adjusting Sleep Patterns Based on Statistical Feedback

If you are working on improving your sleep habits, tracking your sleep duration and quality over several weeks can provide valuable feedback. By analyzing the data statistically, you can identify which factors (e.g., bedtime, screen time before sleep, sleep environment) have the most significant impact on your sleep quality. Statistical significance helps you identify meaningful patterns, allowing you to adjust your sleep routine accordingly.

10.3   Balancing Adaptation with Consistency

Achieving sustained progress requires finding the right balance between adaptation and consistency. While adapting to new information and changing circumstances is essential for improvement, maintaining consistency ensures steady progress toward long-term goals. In this section, we will explore how to balance these two principles to optimize personal development, particularly through habit formation and goal setting. We will also discuss the mathematical models that can help in tracking progress and adapting strategies while maintaining consistent efforts over time.

10.3.1   Understanding the Role of Consistency

Consistency refers to maintaining a steady effort over time, which is crucial for building habits and achieving long-term success. It involves sticking to a plan or strategy even when immediate results are not visible. The power of consistency is rooted in the concept of **compound growth**, where small, incremental improvements accumulate over time to yield significant results. Mathematically, this can be modeled using **exponential growth** functions, where consistent effort leads to exponential improvements.

Example 1: Consistent Effort in Habit Formation

If you consistently practice a skill or behavior each day, even if for a short amount of time, the effect compounds over time. For example, practicing a musical instrument for 30 minutes daily may seem insignificant in the short term, but over several months or years, this consistent practice can lead to significant improvement. This can be represented by the exponential growth formula:

$$P(t) = P_0 \cdot (1+r)^t$$

Where $P(t)$ is the proficiency at time t, P_0 is the initial proficiency, r is the daily improvement rate, and t is the time.

10.3.2   The Need for Adaptation in Personal Growth

Adaptation is necessary when progress plateaus or when external circumstances change. It involves modifying your approach based on new information or feedback. However, constant adaptation without consistency can lead to a lack of focus and scattered results. The key is to adapt when necessary but also stick to fundamental routines that lead to incremental growth. **Statistical models** like **feedback loops** can help guide when an adjustment is needed.

Example 2: Adapting a Workout Plan Based on Progress

If you're training for a marathon, your body may adapt to a certain routine after a few months. At this point, adaptation is required: you might increase the intensity or change your running schedule to prevent a plateau. However, the consistent effort of training, even when adapting the plan, is what leads to overall improvement. The decision to adapt can be informed by tracking performance metrics such as pace, distance, and endurance, using statistical tools to determine when the body has adapted.

10.3.3   Mathematical Models for Balancing Adaptation

The balance between adaptation and consistency can be modeled through **adaptive learning curves**, where the rate of improvement changes over time. A typical approach is to maintain consistent effort while adjusting the parameters as needed. For example, **learning curves** often start with rapid progress and slow down as the skill level increases, requiring more targeted efforts for improvement. By monitoring this curve, individuals can determine when adaptation is needed without sacrificing consistency.

Example 3: Learning Curve Model

The learning curve formula can be applied to many areas of personal growth, where the rate of improvement decreases over time as skills become more refined. The formula is given by:

$$Y(x) = A \cdot x^b$$

Where $Y(x)$ is the cumulative output, A is the starting output, x is the number of repetitions or time spent, and b is the learning rate. The formula shows that initially, the improvement rate is steep, but it slows down over time. This provides insight into when adaptation (e.g., increasing practice or changing techniques) is required to maintain progress.

10.3.4   Striking the Right Balance for Sustained Progress

To achieve sustained progress, the key is to adapt when the current approach is no longer yielding improvements but to remain consistent in executing core habits. Using mathematical models to track and measure progress allows you to identify when adaptation is needed. It is important not to change your strategy too often, as this can disrupt the progress already made. Instead, adapt strategically, ensuring that consistency is maintained over time.

Example 4- Balancing Adaptation with Consistency in Time Management: Suppose you've been using a time management system (such as the Pomodoro technique) for several months and notice that your focus and productivity are declining. This may be a sign that adaptation is needed: you could tweak your Pomodoro intervals or switch to a different technique. However, the key is to keep a consistent daily work schedule, even if the method itself changes.

Chapter 11

Practical Tools and Apps for Tracking and Improving

This chapter focuses on real-world applications of mathematical concepts in improving personal habits, productivity, and overall self-growth. It examines the effectiveness of various apps and tools that help individuals track their progress and make data-driven decisions. This chapter highlights case studies of people successfully using technology to transform their routines and achieve their goals, demonstrating how mathematical techniques, such as data analysis, feedback loops, and habit reinforcement, can enhance personal development.

11.1 Review of Tools and Apps

In today's digital world, numerous tools and applications utilize mathematical concepts to help individuals track and improve their habits and productivity. These tools leverage algorithms, statistics, and data analysis to provide actionable insights and make habit tracking more efficient. In this chapter, we will explore some of the most effective tools and apps that apply mathematical concepts to optimize personal development. From habit tracking apps that quantify progress to productivity tools that visualize trends, these tools integrate mathematical principles to support goal achievement.

11.1.1 Habit Tracking Apps with Quantitative Features

Many habit tracking apps help users maintain consistency by allowing them to track their daily habits and visualize their progress through numerical data. These apps typically employ basic mathematical concepts, such as counting, averaging, and time series analysis, to help users stay on track.

Example 1: Habitica

Habitica is a popular habit tracking app that gamifies the process of habit building. It assigns points to completed tasks and displays progress through visual charts and statistics. Users can track streaks, measure completion rates, and analyze trends over time. The app also uses algorithms to provide feedback on user performance and suggest adjustments based on patterns of success or failure.

Example 2: Streaks

Streaks is another app that helps users track daily habits by focusing on streaks. The app visualizes progress with a simple display of uninterrupted habit days and calculates the longest streak achieved. It also allows users to set goals and track completion rates, which can be analyzed using simple arithmetic and visual aids.

11.1.2 Productivity Tools Using Mathematical Models

Productivity tools help users optimize their time by using mathematical concepts such as task prioritization, time allocation, and efficiency analysis. These tools often provide visualizations that help track and adjust workflows.

Example 3-Todoist: Todoist is a productivity app that helps users organize their tasks using a priority system and deadlines. It utilizes algorithms to calculate task completion rates, time estimations, and project progress. Todoist also uses data analysis to suggest optimal task priorities based on historical performance and task duration, leveraging time series analysis to estimate future productivity.

Example 4: Rescue Time

Rescue Time tracks computer and smartphone usage to analyze how time is spent across different activities. It categorizes activities using predefined labels and calculates the proportion of time spent on productive vs. non-productive tasks. By using statistical analysis, it identifies productivity trends and offers recommendations for time optimization.

11.1.3   Time Management Apps

Time management apps apply mathematical principles, such as scheduling algorithms, to optimize daily routines and improve overall productivity. These apps consider task durations, deadlines, and priorities to recommend an optimal schedule.

Example 5: Trello

Trello is a project management tool that employs a Kanban board system, which visually organizes tasks by status (e.g., "To Do", "In Progress", "Completed"). It helps users manage projects efficiently, using algorithms that automatically prioritize tasks and suggest optimal workflows. Trello's built-in analytics tools allow users to track their progress, providing quantitative metrics such as time spent per task or project completion rate.

Example 6- Clockify: Clockify is a time tracking tool that tracks how time is spent across various tasks. It uses simple mathematical calculations to estimate time spent on each task and generates reports showing total time spent on specific projects. Clockify provides insights into efficiency, helping users adjust their schedules and productivity strategies.

11.1.4   Fitness and Health Apps

Fitness and health apps integrate mathematical principles to track progress toward health and fitness goals. They rely on quantitative data such as calories burned, heart rate, steps taken, and workout performance to assess fitness levels and suggest improvements.

Example 7: MyFitnessPal

MyFitnessPal helps users track their nutrition and exercise habits. It uses mathematical calculations to provide insights into calorie consumption versus expenditure, offering suggestions for improvement based on historical data. By analyzing data trends, the app helps users set realistic health goals, such as weight loss or muscle gain, using regression models to predict outcomes based on previous behaviors.

Example 8: Fitbit

Fitbit is a wearable device that tracks steps, heart rate, sleep patterns, and other fitness metrics. It uses statistical methods to estimate calorie expenditure and adjust recommendations based on past performance. Users can track their progress with visualizations such as step counts and active minutes, and the app's algorithm provides feedback on how to optimize exercise routines for greater fitness gains.

11.1.5   Integrating Mathematical Tools

The power of these apps lies in their ability to combine data collection, mathematical analysis, and real-time feedback to improve habits, productivity, and health. By integrating such tools into your daily routine, you can better understand your behaviors, track progress, and make adjustments based on actionable insights. Combining these apps with mathematical concepts allows you to stay motivated and make informed decisions that lead to sustained personal growth.

Example 9: Using Multiple Apps for a Comprehensive Approach

Many individuals use a combination of habit tracking apps, productivity tools, and fitness trackers to take a holistic approach to personal growth. For example, you could track your exercise routine with Fitbit, use Todoist to manage your work tasks, and use RescueTime to analyze time spent on tasks throughout the day. By combining these apps, users can get a comprehensive view of their progress across different aspects of their lives, making data-driven decisions to optimize their personal development.

11.2   Integrating Tech Solutions into Daily Routines

Incorporating technology into daily routines can significantly enhance personal productivity, habit formation, and overall well-being. By utilizing various tech tools and apps that employ mathematical concepts and data analysis, individuals can make informed decisions, track progress, and stay on course toward their goals. This section explores how to seamlessly integrate these tech solutions into your daily life for optimal results.

11.2.1   Leveraging Technology for Habit Formation

Technology can play a key role in maintaining consistency and tracking progress when building new habits. Habit-tracking apps like Habitica or Streaks provide instant feedback, visual representations of progress, and data-driven insights. These tools help users stay motivated and stick to their routines by making it easier to visualize the effects of their efforts. Integrating these tools into your daily routine can ensure that new habits are consistently practiced and tracked, leading to long-term success.

Example 1: Daily Habit Reminders

Setting up daily reminders on apps like Habitica can help you stay consistent with your habits. For instance, if you're working on a goal like exercising daily, a notification at the same time every day can ensure you don't forget to complete your task. Over time, this increases the likelihood of habit formation by creating a sense of structure and accountability. The mathematical principle behind this is habit reinforcement, where repeated behavior is reinforced through consistent cues, enhancing habit formation.

11.2.2   Maximizing Productivity

Task management tools, such as Todoist, Trello, or Asana, allow you to organize your day, prioritize tasks, and stay on top of deadlines. These tools use algorithms to suggest the most efficient task prioritization based on deadlines, task duration, and complexity. By integrating these tools into your routine, you can streamline your workflow and increase your efficiency. Consistently using task managers helps you stay on track with personal and professional commitments, making it easier to achieve long-term goals.

Example 2: Daily Task Planning

For example, using Todoist's daily planner feature, you can plan your day by breaking tasks into smaller subtasks, setting deadlines, and assigning priorities. Over time, Todoist's algorithm adjusts based on your task completion history, suggesting more efficient ways to organize your work. This use of technology helps optimize time management and ensures important tasks are completed on time, reducing procrastination and increasing overall productivity.

11.2.3 Using Fitness and Health Apps to Improve

Fitness trackers like Fitbit, MyFitnessPal, or Apple Health can be integrated into daily routines to monitor physical activity, nutrition, sleep, and overall health. These apps track key metrics such as steps, calories burned, heart rate, and sleep patterns, providing users with actionable insights into their well-being. By syncing these tools with your daily routine, you can monitor progress toward fitness goals and adjust your activities as needed to stay healthy and productive.

Example 3: Syncing Fitness Data for Health Goals

For instance, using the Fitbit app to monitor daily steps can help you stay motivated to reach your fitness goals. If your goal is to walk 10,000 steps a day, Fitbit will track your progress and provide reminders to move throughout the day. As you adjust your routine, Fitbit's data helps you evaluate performance trends, such as improving your average daily steps over time, which can be mathematically analyzed to visualize progress.

11.2.4   Incorporating Financial Management Tools

Financial management apps, such as Mint or YNAB (You Need A Budget), track income, expenses, savings, and investments. These apps help users maintain budgets, set financial goals, and monitor progress. By integrating these tools into your daily routine, you gain visibility into your financial situation and can make data-driven decisions to achieve financial security and growth.

Example 4: Budget Tracking and Expense Monitoring
Using Mint to track monthly expenses, you can categorize purchases and compare your spending against your budget. By consistently tracking your expenses, Mint provides a data-driven approach to ensure you stay within your financial goals. The app uses statistical analysis to give feedback on your spending habits, helping you adapt your behavior to meet your savings targets.

11.2.5   Balancing Technology Use for Productivity

While integrating technology into daily routines can improve productivity and efficiency, it is important to maintain a balance. Excessive reliance on technology can lead to burnout, distractions, or over-optimization, where you spend more time adjusting and managing tools than achieving your goals. It's essential to use tech solutions as a complement to your personal efforts, not as a substitute for motivation and discipline.

Example 5: Avoiding Overuse of Productivity Tools
While tools like Trello and Todoist can be helpful, constantly adjusting your tasks and checking app notifications can become counterproductive. To avoid overuse, set specific times during the day to check and update your tools, and ensure you prioritize the actual execution of tasks over simply planning or adjusting them. By balancing the use of technology with time dedicated to focused work, you can optimize your productivity without over-relying on tools.

11.3 Case Studies of Successful Habit

In this section, we explore real-world case studies of individuals who successfully implemented mathematical principles, habits, and productivity tools into their daily routines. These case studies demonstrate how the integration of tech solutions, data analysis, and strategic planning can lead to measurable improvements in personal habits, productivity, and overall life goals. By analyzing these examples, we can better understand how to apply the same techniques to our own routines for success.

Case Study 1: Transforming a Daily Routine

John, a software engineer, struggled with maintaining a regular exercise routine while balancing a demanding job. He decided to use the app Habitica to gamify his habit-building process and track his daily workouts. By setting specific goals such as exercising for 30 minutes per day, John created a structure to reinforce his commitment. The app rewarded him with points for every workout completed, and he could visually track his streaks over time.

Outcome:

John saw a 40% improvement in his exercise consistency over a three-month period. The immediate visual feedback and daily reminders helped him maintain his routine, eventually turning exercise into a habit. Using Habitica's goal-setting features, he also tracked his productivity and other personal goals, which encouraged further improvements.

Mathematical Insights:

By applying habit-tracking and feedback loops, John benefited from the mathematical concept of habit reinforcement through repetition. The use of streak tracking provided him with both intrinsic (motivation from achievement) and extrinsic (points and rewards) incentives. This is similar to reinforcing behavior through positive feedback loops, supported by data analysis within the app.

Case Study 2: Optimizing Task Management

Sarah, a marketing manager, faced challenges in organizing her daily tasks and staying on top of multiple projects. She decided to implement a structured task management system using Todoist. Sarah set clear priorities for her tasks, assigned deadlines, and utilized Todoist's built-in productivity analytics to track her progress.

Outcome:

Within six weeks, Sarah improved her task completion rate by 25%. By using Todoist's algorithm to analyze her task history and prioritize based on deadlines, she reduced procrastination and ensured that high-priority tasks were completed on time. Todoist's automatic task prioritization and progress visualizations helped her stay organized and focused on her goals.

Mathematical Insights:

Todoist's task prioritization system used a form of decision tree analysis, where the app analyzed her task history and estimated which tasks would yield the most significant returns if completed first. Sarah was able to apply Pareto's principle (80/20 rule) by focusing on the 20% of tasks that provided 80% of the value. This allowed her to maximize efficiency and productivity.

Case Study 3: Achieving Financial Goals

James, a recent college graduate, struggled with managing his finances after starting his first full-time job. He began using the app Mint to track his income, expenses, and savings goals. Mint provided him with real-time financial reports, categorized his spending, and set up automatic alerts for overspending in certain categories.

Outcome:

Within four months, James managed to save 15% of his monthly income by adhering to a budget set within Mint. The app's data visualization tools helped him better understand his spending patterns, allowing him to make informed decisions about where to cut back and how to prioritize savings. By integrating Mint into his daily routine, James was able to align his finances with his long-term financial goals.

Mathematical Insights:

Mint applied budgeting principles using simple arithmetic and financial modeling to calculate James' monthly savings rate. By analyzing monthly expenses and comparing them to his budget, Mint used statistical analysis to predict whether he was on track to meet his savings goals. This allowed James to adjust his behavior in real-time, based on data-driven insights, improving his financial health.

Case Study 4: Enhancing Health and Fitness

Maria, a busy professional, found it difficult to stay on top of her fitness goals. She decided to use Fitbit to track her daily steps, heart rate, and sleep patterns. With the help of Fitbit's app, she set daily fitness goals, including 10,000 steps per day, and monitored her progress through the app's daily reports.

Outcome:

After three months, Maria achieved a 30% increase in her daily step count and improved her sleep quality by 15%. The feedback loops from Fitbit's daily progress reports kept her motivated, and the app's data allowed her to adjust her activity levels when needed. By integrating fitness data into her daily routine, Maria created a consistent habit of regular physical activity.

Mathematical Insights:

The Fitbit app used basic statistical analysis to calculate Maria's average daily steps and activity levels, as well as to forecast future activity levels based on trends. The app's algorithms helped Maria identify patterns in her behavior, including the times of day she was most active, which allowed her to optimize her fitness schedule for maximum impact.

Case Study 5: Productivity Optimization

David, a freelancer, struggled with time management and often felt overwhelmed by competing deadlines. To gain better control of his schedule, he began using the app Clockify to track how he spent his time on various projects throughout the day. He categorized tasks and tracked his hours, comparing the time spent with the value produced.

Outcome:

David increased his productivity by 20% in two months by optimizing the time he spent on high-value tasks. Clockify's real-time tracking and analysis helped him identify periods of unproductive work, enabling him to eliminate distractions and focus on high-priority tasks. He also used Clockify's reporting feature to analyze his time distribution, adjusting his schedule to maximize output.

Mathematical Insights:

Clockify used time series analysis to track and forecast how time was being allocated across various tasks. By calculating productivity ratios (e.g., value per hour), David was able to identify areas where he could improve efficiency. Clockify's data helped him adopt a more results-oriented approach, focusing on high-return activities and reducing time spent on low-value tasks.

Chapter 12

Conclusion

This chapter summarizes the key principles of the book, emphasizing the importance of integrating mathematical thinking into daily life for personal growth. It encourages ongoing tracking, iteration, and adjustment, while balancing data-driven habits with personal values and flexibility. The chapter concludes that personal development is a continuous, adaptable process where mathematical tools guide intentional, meaningful progress, leading to a balanced and purposeful life.

12.1   Summary of Key Principles from the Book

Throughout this book, we have explored how mathematical principles, statistical models, and data-driven approaches can be applied to enhance various aspects of personal growth, from habits and productivity to financial and health management. Here, we summarize the key principles discussed in each chapter:

1. Quantification for Self-Improvement: The power of measuring, tracking, and quantifying personal habits and progress has been emphasized as the foundation of self-improvement. By breaking down complex behaviors into measurable components, individuals can monitor their performance and make informed adjustments.

2. Habit Formation through Data: Using mathematical models such as exponential growth and compounding effects, we demonstrated how small, incremental improvements lead to significant changes over time. Habit loops, tracked consistently, form the basis of long-lasting positive behaviors.

3.Productivity Optimization: We discussed tools such as the Pareto Principle, Eisenhower Matrix, and productivity ratios to prioritize tasks, eliminate inefficiencies, and focus on high-impact activities, ultimately improving personal and professional productivity.

4. Goal Setting and Tracking: Setting SMART goals and breaking them down into smaller, manageable steps allows for continuous progress. The use of feedback loops and tracking mechanisms helps ensure goals stay relevant and achievable as circumstances change.

5. Financial Management and Planning: Budgeting, saving, and investing are optimized using mathematical models, such as compounding and forecasting. By aligning finances with personal values and tracking expenditures, individuals can achieve greater financial stability and growth.

6. Data-Driven Health and Fitness: Mathematical approaches applied to health data, such as calories burned, heart rate, and exercise performance, provide valuable insights for setting and achieving fitness goals, allowing individuals to monitor their progress effectively.

7. Resilience and Learning from Failure: Using probability, growth mindsets, and statistical analysis, we discussed how setbacks can be modeled, analyzed, and used as opportunities for learning and improvement, ultimately building resilience and fostering a continuous growth mindset.

12.2 Mathematical Tools for a Balanced Life

By integrating these mathematical principles into daily routines, individuals can create personalized systems to track progress, make informed decisions, and achieve a balanced life. The key is to use data and feedback loops as tools to enhance self-awareness and guide personal development. This approach leads to greater consistency, focus, and, ultimately, success in various areas of life.

12.3   Encouragement for Ongoing Tracking

As we conclude this journey through the mathematics of habits and productivity, it is crucial to recognize that optimization is not a one-time achievement but an ongoing process. Continuous tracking, iteration, and adjustment are essential components of personal growth. The process of tracking habits, goals, and productivity is dynamic and should evolve based on feedback, results, and changing circumstances.

Tracking: Regularly measuring progress ensures that you stay on track, helping to identify areas where adjustments may be needed. The use of tools and metrics to track your habits and productivity can provide real-time insights into your performance.

Iteration: Personal growth requires constant refinement. Small changes made regularly, based on what you track, can lead to significant improvements over time. Be willing to adjust your methods as you learn what works best for you.

Adjustment: The ability to adapt and adjust is critical to long-term success. Reassessing goals, revising strategies, and staying flexible in response to new challenges or information ensures that you continue moving toward your objectives, even when circumstances change.

Remember, progress is often incremental. Even when things don't go as planned, it's the consistency of small improvements and adjustments that leads to lasting success. Embrace the iterative process, track your progress, and remain open to change to keep optimizing your journey toward personal growth.

12.4 Final Thoughts on Balancing Data-Driven Habits

While mathematical tools, data-driven habits, and productivity models provide powerful ways to optimize your personal growth, it is essential to remember that they are not ends in themselves but means to an end. The ultimate goal is to align your actions with your personal values, priorities, and well-being.

Aligning with Personal Values: Numbers and data should always serve as tools that help you live in accordance with your values. Whether it's managing time, setting goals, or tracking habits, it's important to ensure that these actions support what truly matters to you. A data-driven approach should never compromise personal well-being or what you stand for.

Embracing Flexibility: Life is unpredictable, and there will be times when strict adherence to a plan or metric isn't realistic. Flexibility is key in adapting to changes while maintaining progress. Data-driven habits should allow for adjustments based on circumstances, new insights, or personal shifts in priorities.

Balance Between Rigor and Adaptability: The best approach is a balanced one—rigorous enough to track progress and create measurable outcomes, yet adaptable enough to allow for spontaneity, creativity, and personal growth. The most successful strategies will incorporate both data-driven precision and room for personal discretion.

In conclusion, a thoughtful balance between data-driven habits and flexibility will lead to a fulfilling and sustainable path toward personal growth. Use numbers and data to guide your actions, but always ensure that they align with your deeper sense of purpose, values, and the flexibility to adapt as life evolves.

www.ingramcontent.com/pod-product-compliance
Lightning Source LLC
Chambersburg PA
CBHW071519220526
45472CB00003B/1078